Praise for *The Brain Is Wider than the Sky*

'As readers have come to expect from Bryan Appleyard, his new book is another literate and sensitive reflection on how science is changing our self-understanding'   Steve Fuller, *Literary Review*

'An acerbic exposé of the empty promise of the computer age'
Michael Burleigh, James McConnachie, *Sunday Times*

'Brian Appleyard's *The Brain Is Wider than the Sky* is a beautifully written defence of human complexity in the face of the corporate mechanisation of our lives. If you are frustrated by automated queuing, this is one for you'   Michael Burleigh, *Sunday Telegraph*

'Appleyard is scientifically literate, vigorous and intelligent . . . essential reading'   Simon Ings, *Observer*

'Bryan Appleyard is our foremost guide to understanding contemporary culture. This exploration of what it means to be human today grips the reader from the first page'   John Gray, author of *Straw Dogs*

'There are great science writers and there are great arts writers – and then there's Bryan Appleyard. He's both'   John Humphrys

'Bryan Appleyard is that rarest of rare birds, a journalist who can mine factual subjects for their poetic resonance right across the spectrum. He is our main man for this kind of writing'   Clive James

'Appleyard is a gifted writer, able to explain both the beauty of a Hockney drawing and the mathematical unit used to measure how many computations processors like our brains are capable of performing . . . it's always fascinating, and always clearly expressed'
Helen Lewis-Hasteley, *New Statesman*

Bryan Appleyard is an award-winning special features writer and columnist for the *Sunday Times*. He is the author of several books, including *Understanding the Present*, *Brave New Worlds: Genetics and the Human Experience* and *How to Live Forever or Die Trying*. He lives in London.

www.bryanappleyard.com

By Bryan Appleyard

The Culture Club: Crisis in the Arts
Richard Rogers: A Biography
The Pleasures of Peace: Art and Imagination in Post-War Britain
Understanding the Present: Science and the Soul of Modern Man
The First Church of the New Millennium: A Novel
Brave New Worlds: Genetics and the Human Experience
Aliens: Why They Are Here
How to Live Forever or Die Trying: On the New Immortality
The Brain Is Wider than the Sky: Why Simple Solutions Don't Work
in a Complex World

The Brain is Wider than the Sky
Why Simple Solutions Don't work in a complex world

BRYAN APPLEYARD

PHOENIX

A PHOENIX PAPERBACK

First published in Great Britain in 2011
by Weidenfeld & Nicolson
This paperback edition published in 2012
by Phoenix,
an imprint of Orion Books Ltd,
Orion House, 5 Upper St Martin's Lane,
London WC2H 9EA

An Hachette UK company

3 5 7 9 10 8 6 4

A CIP catalogue record for this book
is available from the British Library.

ISBN 978-1-7802-2015-4

Typeset by Input Data Services Ltd, Bridgwater, Somerset

Printed and bound by CPI Group (UK) Ltd, Croydon, CR0 4YY

The Orion Publishing Group's policy is to use papers that
are natural, renewable and recyclable products and
made from wood grown in sustainable forests. The logging
and manufacturing processes are expected to conform to
the environmental regulations of the country of origin.

www.orionbooks.co.uk

For Richard Appleyard

CONTENTS

INTRODUCTION

This is a book about, in roughly this order, neuroscience, machines and art. It began when, in August 1994, I visited Microsoft in Seattle and spent a couple of hours with the company's co-founder and then chief executive officer, Bill Gates. In the course of the visit, something began to form in my mind. It was too vague to be called a thought; rather, it was a mood, an anxiety, an uncertainty, a riddle, but it seemed to me, even in my vagueness, to be fundamental to the nature of the new world that was then just being born and in which we now live.

Over the ensuing years I explored this mood of mine, either deliberately or, because of some unexpectedly interesting book, article or meeting, accidentally. A series of clear polarities began to emerge from my initial confusion: mind and machine, art and technology, real and virtual and, most consistently, complexity and simplicity. This book is my attempt to explain and, perhaps, justify what I first felt in 1994. It wanders over many disparate fields – from poetry to neuroscience, from computer games to finance, from philosophy to climate change, from iPhone apps to iPad art – but they are all tied together by my attempt to understand the true nature of the vast change that has happened in the last two decades. It is an attempt to identify what I believe is the primary dynamic of the modern world.

Before I return to that visit to Microsoft, first, I want to give some idea of the broad themes of this story.

Some time between 1858 and 1865, her most creative period, Emily Dickinson sat at the eighteen-inch-square writing table in her bedroom and composed a twelve-line poem. This was in Amherst, Massachusetts, where Dickinson was known as a brilliant, caustic, witty but odd member of one of the small town's most prominent families. She spent much of the fifty-five years of her life in seclusion. Her recent biographer, Lyndall Gordon, suggests this was because she suffered from epilepsy, a condition that may have been considered shameful for a woman of that time. Nevertheless, locked away in Amherst and in her head, Dickinson wrote nearly 1,800 poems that, long after her death, led to her being considered by many to be America's greatest poet.

This particular poem has three verses of four lines. The lines are characteristically short – alternately eight and six syllables – and punctuated by dashes, her usual style. The tone, rhymes and form seem light but the lines are heavy with meaning. The first verse alone anticipates much contemporary scientific and philosophical thought:

The Brain – is wider than the Sky –
For – put them side by side –
The one the other will contain
With ease – and You – beside –

Dickinson, for all her delicate seclusion, was an intellectually robust individualist, often riskily at odds with the puritanical culture of Massachusetts. She read the philosopher Ralph Waldo Emerson, one of the most celebrated figures of the age, and took from him a belief in the power of the imagination to transcend the limitations of the human condition. This points to the first and most obvious meaning of that verse: it is a celebration of the sheer scale of the brain, its ability to encompass the entire world – a poignant but potent idea for a little-travelled woman sequestered in her bedroom in Amherst.

But there is a twist in that verse. Not only does the brain include the whole sky, it also includes the mind – the 'You' or self that perceives the sky and can think about it. It is this mind that makes the brain wider than everything. Dickinson could have replaced 'sky' with 'universe' in that first line and it would still be true that the brain was wider because there is something in the brain that is not included in the physical world. We contain the sky 'with ease' because we have infinitely capacious minds.

So the mind includes but does not seem to belong to the physical universe. How can this be? This is what is known in philosophy as the 'hard problem of consciousness'. How does consciousness arise from the ordinary physical matter – mainly fat and water – of the brain? We have, for the moment, neither a philosophical nor a scientific answer to this. Human consciousness remains a mystery, the final wonder of the world, the earth's last wilderness. Great art, like the poetry of Emily Dickinson, affords us glimpses not of an answer but of the deep truth and ultimate difficulty of the hard problem.

Dickinson would have known little of science and, of course, nothing of the science of our time. But epilepsy and the way in which she turned her unique sensibility into art made Dickinson's poetry anticipate, with uncanny accuracy, the most urgent and fundamental preoccupations of our age, and the questions raised by contemporary neuroscience, our most distinctive scientific project. 'If,' Lyndall Gordon writes, 'the twenty-first century is to explore unknown pathways of the brain, Dickinson's poetry is replete with information about dysfunction and recovery.'

Modern science is the exploration of the physical world through observation, experimentation and, latterly, through computer modelling. Since Galileo first looked through a telescope in 1609, the insights of science and the fruits of those insights in technology have been astounding. We can now explore the ultimate constituents of matter and the outer limits of space and time. We can catalogue the human genome, detect and even describe planets circling stars thousands of light years away. We have eradicated or suppressed lethal diseases like smallpox and polio and, in the developed world, we have almost doubled human life expectancy in the last 150 years. We have connected the world with the single communications system of the internet, men have walked on the moon and our machines have trundled across Mars. In spite of which, we can still only stare in wonder and ignorance at the human mind that has achieved these things. The brain has remained much wider than the sky.

But, at last, we may have a way into the brain and the mind. Thanks, primarily, to Functional Magnetic Resonance Imaging (fMRI), we can now watch the brain in action. The fMRI machine detects blood flow in the brain – increased blood flow in a specific area is associated with increased brain activity in that area. So now scientists have a tool with which they can correlate mind processes with clearly identifiable brain areas and processes. They have constructed, it seems, at least the foundations of a bridge connecting the mind and the brain and therefore, potentially, a way of solving the hard problem.

The possibility has inspired both awe and investment. Neuroscience, thanks to the MRI machine, is now the most fashionable science. Neuroscience stories routinely make it into the newspapers. The latest speculations are reported as if they are fully established facts like the discovery of a new continent or planet. MRI scans have become scientifically underpinned horoscopes, new ways of understanding ourselves and our destinies.

Results of brain scans are being studied for insights into economics, politics, advertising, marketing, sociology, anthropology, religion and art. Politicians think neuroscience will help them know their voters better and marketers believe it will help them sell more to their customers. They no longer have to rely on the vagaries of surveys, focus groups and responses to questionnaires; they can, instead, watch what is really going on in the minds of their targets. With the ever-increasing power and precision of the scanners, it seems there will be no limit to our ability to know the workings of the human mind.

The problem is that the human brain may be the most complex thing so far discovered in the physical universe and, as any good neuroscientist will admit, it is far from clear what we are seeing when we look at fMRI scans. Are the increased blood flows causes or effects of the activities of the mind? Do they merely demonstrate the blindingly obvious – that *something* is happening in the brain when we think? Furthermore, the language in which we describe the results is confusing. Neuroscientists often say, for example, 'You are accessing your language centres.' But who and where is this 'you'? It is certainly not yet detectable by the machine. Rather than solving the hard problem, the fMRI machine is just restating it.

But, to the thoroughgoing materialist, the hard problem *must* be soluble. If the mind is not explicable in purely physical terms, then what is, for us, the most important thing in the universe, the thing that seems to *create* the universe every day, lies beyond materialism. We might have to say the mind may still be material but it is incapable of understanding itself. Perhaps asking our scanners to scan the mind is like asking a camera to photograph the film or sensor with which it takes pictures. It cannot, logically, be done. In this interpretation, materialism survives but only just and on very shaky foundations.

On the other hand, if the materialist faith is true, and if these logical puzzles can be made to disappear, then the brain is a machine like any other. It can be understood as a machine, fixed as a machine and replicated like a machine. Indeed, entirely new types of brain can be created made of silicon – or some other material – rather than fat and water. The inconveniences of the biological can be superseded by the comforts of the machine.

This may happen in the near future. The current rate of tech-nological innovation has convinced some that we are within two or three decades of building the ultimate machine. At this moment – known by enthusiasts as the Singularity – we will build a computer more intelligent than ourselves which is capable of either booting itself into ever higher levels of intelligence or of building new, even smarter machines. It will relieve us of all our technical and biological burdens. The machine will take over.

This idea of the thinking machine, though now very familiar, is actually quite new. The meaning of the word 'machine' has changed radically over the last twenty years or so. The first great machine of the Industrial Revolution was the steam engine and it remains, in our imaginations, the symbolic machine – big, metal, noisy, often dirty and, in fact, rather clumsy. It demonstrated the effort and contrivance needed to replace or improve on nature, primarily, in this case, horses.

The first computer, built, though never completed, by Charles Babbage in the nineteenth century, looked like any other machine of the Industrial Revolution. At first sight, it was just a rather special kind of engine and, indeed, Babbage called it a Difference Engine. His later version, called the Analytical Engine, was, unlike the first, programmable using punched cards. The appearance of these machines, combined with the fact that they were both called 'engines', disguised the truth that they were entirely new machines. Looms produced cloth, steam engines pulled wagons and guns killed people – they changed the physical, visible world. But Babbage's computers solved sums. Their output was entirely abstract, pure thought. Physically, they did not change the world at all; mentally, however, they were the start of a revolution that would change everything.

This moment of transition from material to abstract machine output was the moment when our new machine age was born. The virtual world, in which we now partially live, was produced by machines that are the descendants of Babbage's engines. Most importantly, it was the moment when a new model of the human mind came into being, a machine model. After Babbage, it became commonplace to imagine machines thinking.

But the machines in our everyday life did not really start changing until the arrival of the personal computer. Before that we had, for example, cars and washing machines, essentially industrial-age engines. Radios and TVs were different. They plucked invisible waves from the air and made sounds and pictures we could hear and see. But that is *all* they did. We, meaning the lay consumer, could not make them do something else.

The computer, however, is a *universal machine*. That term is important because it defines where we are now and where we may be in the near future. The structure of the modern computer was first outlined by the great British mathematician Alan Turing in 1937. It was called the Universal Turing Machine. This meant that it could calculate anything that could be calculated. The intellectual implications of this idea were enormous – not least, there was the issue of what could *not* be calculated – and so were the practical ones. Turing's machine was the computer as we now know it.

The arrival of the personal computer brought this machine into the home. We were given a universal machine to go with the car, the washing machine and the TV. At first, however, its universality was baffling. Consumers were not programmers and, for most of us, computers were simply glorified typewriters, or word processors as we now called them. Or, from 1979 with the arrival of the Visicalc spreadsheet software, they helped businesses keep track of their accounts. Or they played games.

Somehow, it wasn't enough. In an episode of the TV comedy *Friends* from the mid-1990s, Chandler opens a new laptop and brags about its apparently enormous capabilities. Yet, when asked what he would do with it, he sheepishly replies that he might play a few games. In fact, as I shall explain later, games were to prove a crucial driver in the explosion of computing power that was to follow. But, at the time, games were an explicit acknowledgement that we had no idea what to do with these potent and mysterious new machines.

But the internet made the true universality of the computer evident. Through sudden access to what seemed like all the information in the world, through shopping, social networks, blogs, maps and satellite photographs of the entire planet, through virtual online worlds, we learned that these machines in our homes could do almost anything. And the frantic pace of the increases in processor speed and memory convinced us that, in time, we could drop that 'almost'.

These new machines close in on our lives just as the science based on the most important new machine – the computer – closes in on our minds. Brain-scanning technology is only made possible by the exponential increases in computing power and speed we have seen over the last sixty years. Caught in this pincer movement is the as yet inexplicable human mind, apparently the product of the human brain, still irreducibly complex.

But what does complexity mean? Its definition is precise and is the opposite of the simplicity that has, for more than seven hundred years, been the Holy Grail of Western rationality.

William of Ockham was born in about 1288. Ockham, the place of his birth, is a small Surrey village south-west of London, now about half a mile from the M25 motorway; in William's time it would have been a day's ride from London. He was evidently a bright child and was sent into the city to join and be educated by the Franciscans. He later studied theology, probably at Oxford, though he never completed his course, perhaps because his prodigious intellect was already beginning to lead him astray. He would eventually be excommunicated by the Church after concluding that Pope John XXII was a heretic. He died in Munich aged sixty.

William is now regarded as one of the giants of medieval thought, an equal of Thomas Aquinas or Duns Scotus. Like them, he was a member of a movement known as scholasticism. This was not a single philosophy but a rigorous, analytical and questioning way of thinking that, in many ways, anticipated the scientific method and, therefore, the modern world. He is famous, far beyond the confines of academia, for one big idea which became known as Ockham's razor. At its simplest, this states that 'entities must not be multiplied beyond necessity'. The technical term is 'ontological parsimony', meaning that explanations of the nature of being (ontology) should be as simple as possible (parsimony). Like a razor it shaves away all unnecessary factors. The general principle is that if part of an explanation is not necessary, then it should be removed.

This seems reasonable but there is a problem. William's razor is a logical tool in philosophy, but in science it can only ever be what is known as a heuristic, a method of understanding. In other words, it may be sensible for a scientist to use the razor and go for the simplest explanation, but there is no logical basis for him then to insist that this explanation is correct. However much we may like the idea, we have to accept that the universe has no obligation to be the simplest possible universe. Einstein understood this. He said, 'Everything should be kept as simple as possible, but no simpler.' Yet William's belief in the simplest possible explanation is constantly evoked in the literature of popular science. He has become a secular saint of materialist rationality.

Simplicity has now become a brand. In engineering and design, for example, there is the KISS – keep it simple, stupid – principle, first named by Kelly Johnson of Lockheed Skunk Works. Many architects, especially the designers of urban apartments, still adhere to the command of Mies van der Rohe, 'less is more', on the basis that any unnecessary elaboration is always a bad thing. The media loves simplicity and always shaves news stories, especially political ones, down to the simplest possible two-way arguments. Drugs companies want us *simply* to take a pill. Advertisers live by simplicity because they want people to think that *simply* by buying their product you can solve your problem. Politicians pursue simplicity because they think anything else would distract people from simply voting for them.

Then there is the seductive idea of the simple life. Pressured by their work and the complications of the modern world, people seek simplifications. Most holidays are sold as simple interludes of sun, sand and sea. Specialised holidays, usually with quasi-religious overtones, are 'retreats' from the complications and turmoil of our lives. Organic foods are sold as simplifications freed of all the suspicious interferences of chemicals and genetic manipulation. Alternative remedies are sold as simpler, more natural solutions than the sinister products of Big Pharma. Periodically, simplification fads emerge on the chat shows and daytime TV, led by books with titles like *Simplify Your Life*, *Simplify Your Time* and *Don't Sweat the Small Stuff*.

The simplicity heuristic has escaped from science and entered the human world. Ironically, it is often used, as in those quasi-religious retreats, as a way of escaping from the technological fruits of science. But can it really work in the human world if it is only, in reality, a useful but specialised and provisional technique in science? Can simple solutions work in a complex world?

Furthermore, is this spiritual ideal of simplicity really what it seems? In *Four Quartets* T. S. Eliot speaks of 'a condition of complete simplicity' but then he adds in brackets 'costing not less than everything'. In other words, true spiritual simplicity is, in fact, everything, it is complex. This is something to be achieved, not a default condition to which we can revert like 'pure' food or the simple life.

The opposite of the kind of simplicity that is now so often evoked as a good thing is complexity. An important verbal distinction needs to be made here. For my purposes, complexity is not the same as complication. The two are often used as if they were interchangeable. Complication is merely quantitative: it arises because of the sheer number of elements involved. Complexity is qualitative: something new arises from the various elements

The difference between complicated and complex things is obvious once you look for it. The instruction manuals for certain

Japanese cars and cameras are complicated; they form a disorderly pile rather than a coherent view of the machine and how it works. Computer software – especially Microsoft Word – has grown more complicated over time because faster processors and larger memories have encouraged designers to add more features. The plots of bad movies or novels are often complicated; they do not hang together.

In contrast, a complex thing would be the experience of driving the car, writing a great book with Word or watching *The Man Who Shot Liberty Valance*. In these cases, it all hangs together, something new has emerged that could have happened in no other way and which may feel simple but isn't. It feels simple because complexity comes naturally to us. We are complex and we welcome complexity; it feels like home.

Since Galileo, it has taken four hundred years for science to embrace the idea of complexity. In each of the last three decades of the twentieth century, a new fundamental theory emerged which seemed to modify the view of science in the tradition of William of Ockham and Galileo, a tradition that viewed the world as a coherent and decipherable system of cause and effect.

These new theories were closely linked. In the 1970s it was catastrophe theory, which concerned the way small changes in systems could lead to very large effects. For example, a ship in a storm may rock for a long time and then suddenly capsize. What exactly changed? Chaos theory in the 1980s was similar but went further. Small changes in initial conditions could produce enormous and unexpected effects – the flap of a butterfly's wings in Tokyo causing a storm in New York was the famous example. The crucial point about chaos was that its exact effects from moment to moment were intrinsically unpredictable. This suggested there was a limit to our knowledge of the world.

Then, in the 1990s, there was complexity science. This is often confused with chaos theory, but in fact the two are quite different. Chaos theory is deterministic. Though exact outcomes are unpredictable without infinitely precise knowledge of the initial conditions – exactly how and where the butterfly flapped its wings, what the weather was like, at what angle the wings flap – we do know the range of possibilities of the initial conditions, the 'state space' as it is called. But in a truly complex system like the living world, we cannot even define the state space: complexity is non-deterministic.

As Stuart Kauffman, one of the most prominent complexity theorists, points out, chaos theory was not truly radical; it did not break the Galilean model of science. But true complexity is a revolution in that it reveals a universe in which some aspects cannot be captured by any laws of nature. 'We have entered,' he writes, 'an entirely new philosophic and scientific worldview ... This is, indeed, radical. From it spreads a vast new freedom, partially lawless, in the evolution of the biosphere, economy and human history in this nonergodic [never returning to the same state] universe.'

Many disagree with Kauffman's radical views of the importance of complexity science and some say he has strayed outside the boundaries of 'real' science. I have heard him called 'a poet of science' rather than a scientist. In addition, as a result of uncertainties about what it means and how it might be applied, complexity's exact status as a science has changed repeatedly. There is no doubt that it is still struggling to define itself as a distinct discipline. But, equally, there can be no doubt that complexity science's one big central idea does raise fundamental doubts not just about Galilean science but also about how we see ourselves and where we think we are going.

The first and still the most influential contemporary expression of the basic truth of complexity was in an article published on 4 August 1972 in the journal *Science*. The author was Philip Warren Anderson, a physicist who, in 1977, won the Nobel Prize. His paper, 'More Is Different: Broken Symmetry and the Nature of the Hierarchical Structure of Science', was an attack on reductionism.

Galilean science is based on a belief in reductionism, that, in the words of Steven Weinberg, another Nobel Prize-winning physicist, 'the explanatory arrows always point downwards'. This means that we understand the world by breaking it down into its smallest constituents. In some sense, this computer is 'really' atoms, protons and quarks. Reductionism is the belief that if we understand these fundamental units of matter, then we can ultimately understand everything.

Anderson was writing at a time when reductionism was accepted without question; to a large extent it still is. His key point was that reductionism could not explain how we got from there – subatomic particles, fundamental forces – to here – the world as it is and as we experience it. The explanatory arrows pointed downwards but did they actually explain anything? Plainly not, because we could not use them to deduce causal arrows going upwards.

'The main fallacy,' Anderson wrote, 'in this kind of thinking is that the reductionist hypothesis does not by any means imply a "constructionist" one: the ability to reduce everything to simple fundamental laws does not imply the ability to start from those laws and reconstruct the universe. In fact, the more the elementary particle physicists tell us about the nature of the fundamental laws, the less relevance they seem to have to the very real problems of the rest of science, much less to those of society.'

Reductionism suggests there is a way of looking at the world that must, logically, have greater truth value than any other. This table, says the reductionist, is 'really' composed of almost empty space. Weinberg's arrows all go downward because they are the only arrows we have. We think the laws to which these arrows point 'explain' the universe from the Big Bang to this table, but we cannot show how that might happen. The reductionist would respond that, in theory and, possibly, in the future, we can. The response of Anderson – and the ensuing generation of complexity scientists – would be that, in fact, we can't. The upward causal arrows, as opposed to the downward explanatory ones, are unknowable.

So, says Anderson, the reductionist idea that there was a hierarchy of sciences that always ended in physics was simply wrong: 'Psychology is not applied biology, nor is biology applied chemistry.' With a graceful literary flourish, he ended his paper with a famous dialogue that took place in Paris in the 1920s.

F. SCOTT FITZGERALD: The rich are different from us.
ERNEST HEMINGWAY: Yes, they have more money.

As well as being the greater artist, Fitzgerald was also the greater scientist. Reductively counting dollar bills is no way to understand the full complexity of what wealth does to a person. The rich, Fitzgerald was saying, are different in kind not just in quantity. More, as Anderson would put it, is not just more, it is different.

Anderson and the complexity scientists were reiterating the ancient insight that some things are greater than the sum of their parts. At one level this may seem to be nothing more than common sense. We know that people are more than their liver, lungs and other organs, and we certainly know they are more than their atoms and molecules. But what common sense misses is the deep significance of the word 'more'. If a system like a human body produces something more than its parts, where does that 'more' come from and what is it? The question is fundamental and threatening to reductionism and simple-minded materialism.

In complex systems 'more' is defined as 'emergent properties'. These are things the system can do that are not predictable from the constituents of that system, because they do not arise simply as a sum of all the properties of the parts of the system. A computer may be predictable from parts such as its memory, processor, hard disc and screen; it is emphatically not predictable from the qualities of silicon, aluminium or glass. But the best example is the most complex system of all, the human brain. Somehow, this particular organisation of fat and water generates the conscious human mind. Cars, wars, office blocks and poetry are all emergent properties of fat and water.

This is a basic summary of a vast and ever-changing scientific discipline, but it is enough, for the moment, to demonstrate the importance of the idea of complexity. Emergent properties may be obvious to the layman but they represent a fundamental change in our understanding of the world. As with chaos theory, complexity theory suggests there is an absolute limit to our knowledge. Emergent properties are not predictable. They are new things in the world that seem to come from nowhere. We can make an attempt to understand them backwards – by tracing the pathways that made them emerge – but this is not necessarily possible even with apparently straightforward complex systems. You can trace the making of a car backwards fairly easily if you just go back to the factory, but what about the history, economics, sociology and chemistry which also went into making that car? Even if you could do that, you still couldn't trace the pathways forwards from, say, the Big Bang with which our universe began to a Porsche Boxster.

Complexity theory has been a revolutionary force. In sociology, anthropology, psychology, political science, finance, management, economics and many other areas, the implications of complexity are slowly being worked out. Implicit in all this is the realisation that complexity is a desirable thing; it stabilises systems and makes them robust, able to survive shocks.

These are unconditionally good developments which mean that, in these areas, we are moving on from some simple-minded thinking. For example, in economics the idea of humans as simple terms in an equation, units like atoms that respond predictably to inputs but with certain outputs, has now been more or less thoroughly discredited. Or, in anthropology, we are now much more aware of the role of entire cultures rather than just the parts that we can fit into our pre-existing models. In politics both the big simplifications of collectivism driven by an all-powerful state and individualism driven by an economic war of all against all no longer have any useful meaning.

Yet traditional Ockhamist–Galilean science has been the most successful human project we have ever known. It has given us wealth, health and steady economic growth. What could possibly be wrong with showering us with all the things we want when we want them? We have grown rich; we are, in our homelands at least, at peace. If the West's 2,500 years of intellectual and physical struggle were not meant to achieve this, what were they meant to achieve? Is becoming more like a machine or, indeed, submitting to a machine so bad if it means we live longer, do not die violently or starve? We are, as a species, ever more effective at transforming the world and our lives.

Since the late twentieth century, however, we have begun to show signs of being all too effective. Nuclear weapons have given us the power to destroy all life or at least our civilisation. Population growth suggests we might strip the planet of its resources. Pollution, primarily through carbon dioxide emissions, threatens to cause uncontrollable heating of the atmosphere. The apocalypse, it seems, is the price we must pay for the triumphs of Galilean science. Perhaps it is time to move on, to think differently.

There are further questions raised by simplification and the rise of the machines: what are we in danger of losing? If we go on down the road to the Singularity, when the machine finally takes over, who will we then be? Are we in danger of losing complexity, the very thing we should value most and the very thing that best defines our humanity? Machines simplify, they categorise and index. They demand simpler forms of behaviour. Like Ockham's razor, they strip us to our essentials.

We are at the beginning of this second machine age. The change involved is so vast we can barely see it. This book is an attempt to describe that change. The first machine age began in the mid-eighteenth century and gave the developed nations 250 years of relentless economic growth. The second gives us – what exactly?

Art has always been the best guide to what we most profoundly are and to what we are becoming. Writers, architects, visual artists and musicians define our species and our predicament. This book starts and ends with that conviction. Poets have seen what is at stake in the two machine ages. Confronted with the clanking, metal machines of the first age, one poet in particular was alarmed at what was being lost:

> The world is too much with us; late and soon,
> Getting and spending, we lay waste our powers;
> Little we see in Nature that is ours;
> We have given our hearts away, a sordid boon!

That poet was William Wordsworth, longing for a return to an enchanted age of gods and myths. The second age has already gone even further than the first in accepting science's two escaped heuristics – Ockham's razor and reductionism – into the human realm. Now we routinely accept the 'nothing but' of simplification and reduction. This table is 'nothing but' empty space, humans are 'nothing but' machines, I am 'nothing but' an animal.

This leaves the poets pining not just for the old gods, but also for our selves. The American poet John Ashbery wrote in 1998, 'Renewed by everything, I thought I was a ghost'. He felt the terrible power of all the novelty of the consuming, machine-using contemporary world, the way everything seems to change all the time, our selves included. The poet's self had become a ghost haunting the shopping malls of the world.

But it is Emily Dickinson's poem 'The Brain is wider than the Sky' that most exactly captures the contemporary sense of the strangeness of human consciousness. It also defends that strangeness, that ability, unique in nature, to include the whole world and the thinking self in the as yet unobserved and undetected location we call the mind. If she were alive now she would be asking: will the machine brain – or our brains when adapted to the demands of the machine – still be wider than the sky?

This book is about neuro- and computer science, about gadgets and games and about some of the great delusions of our time. It is also about what lies above and beyond all of those thing and it begins in Seattle in 1994.

1

A DIVIDED MAN

At the climax of my visit to Microsoft I was left alone in a room.

'You'll want to collect your thoughts,' said the woman from public relations.

Perhaps, I thought, this was a room specially set aside for thought collecting prior to a meeting with Bill Gates.

I had spent a couple of days on the Microsoft campus in Redmond, a suburb of Seattle. This place, I had realised, foreshadowed a new world. Old world companies had offices and factories; new world companies had campuses. The word 'campus' was being used to make it clear that this was not any old drab workplace. A campus, after all, is a kind of home where whole, serious, thoughtful lives are lived.

But the word was not just symbolic, it also signalled a real change. The office – factory distinction had become quaint or meaningless. Computer software is not made of metal or plastic, the sorts of things used in factories, it is made of numbers. Software could be produced in what had been called offices and were now to be called campuses. And software was rapidly becoming the most important thing in the world. It told the hardware what to do; it was the mind of the new machines.

Microsoft's campus was a landscape of trees punctuated by low buildings with ribbons of tinted windows and white walls. Each building was surrounded by parked cars. Signs were few and the tinting of the window glass made the place feel anonymous and, at first sight, deserted, possibly abandoned. Everything was happening inside. In the middle of the campus there was a little pond which they called Lake Bill. It was inhabited by Canada geese that deposited bluish dung all over the place.

When I first arrived, the receptionist did not greet me. She just waited for me to type my name and business into a laptop computer that printed out my pass and told her who to call. Then she said: 'Take a seat.'

Inside each building – they were called Building One, Building Two, etc. – there were just grey corridors lined with small cubicles. In each a person was working at one or two computer screens. A few looked up as I passed, but most were too immersed in their screens to notice anything.

In 1994 the old world was heavily pregnant with the new one. The birth was said to be imminent. Personal computers (IBMs and Apples) had been around since the beginning of the 1980s and I was already an 'early adopter', one of those people who must have new gadgets as soon as they come out.

I had been disappointed. Nothing was connected to anything else, at least not in a useable or useful way. The new machines were only what was inside them. So computers just sat there being more troublesome and demanding than the typewriters they replaced. The hottest new technology was CD-Rom which did, at least, offer us some chance to make these beige boxes do something interesting. Nevertheless, I was excited. We were about to cross a threshold, and in the very near future, all the computers in the world would be connected. The beige boxes would become not just what was inside them but what flowed through them and what flowed through them would be everything. We would become like the all-seeing eye on the dollar bill; nothing would be hidden from us.

This was why Bill had agreed to meet me. He had written a book, *The Road Ahead*, about something he called the Information Superhighway. We would all have access to all the information in the world. This was thrilling. But my meeting was not just about

Bill's book. Microsoft was also about to launch a new operating system codenamed Chicago.

First, they wanted to make sure I understood. I had to spend three days at the campus, being shown round and meeting important people. This was very demanding because I didn't actually know very much about computers.

I met Nathan Myhrvold who was in charge of advanced technology. He was a bearded, chubby man who laughed a lot, mainly at his own jokes. He talked in strange swoops and screeches and seemed dazzled by his own cleverness: 'We're surfing on some great wave of technology and there's this guy on the surfboard [Gates] who can surf.'

There was Chris Peters, in charge of word processing. He looked like a nervous undergraduate, possibly, I imagined, fond of hiking. Unlike Myhrvold, he was not pleased with his cleverness; rather, he was almost oppressively aware of its oddity and limitations. 'We don't hire well-rounded people here,' he said, 'we don't look for well-roundedness in people, we look for computer freakishness. We don't care if you shave or what you wear.'

I asked him about his boss, Bill, and the tentativeness vanished to be replaced by the confidence of the enthusiastic fan. 'He is incredibly important, one of the smartest people of the twentieth century. He has a historical level of genius and he's the most competitive person I have ever met.'

On the advanced consumer side there was Greg Riker who spoke of a 'different species' emerging from all this new technology. He was working on a device he called the 'wallet PC'. Striding around the campus, he seemed, more than anybody else, to embody a new way of life.

In 1994, all the technology had not happened yet; it would not be in our homes and offices for years. This was not surprising; high-tech companies are all racing into the future so, in a sense, everything is always just out of reach. I was, however, in a receptive frame of mind and I happily assumed that whatever Microsoft said was going to happen would, indeed, happen. Having surpassed IBM, the company was, after all, the biggest player in the computer game. Steve Jobs had left Apple in 1985 and the company had lost its hippie, alternative sheen. Microsoft may have been boring in comparison, but at that moment it owned the future.

Even before I met Bill, the visit had begun to make me question my early adopting excitement. I had started to meditate on this new relationship between humans and machines. Everything on the campus suddenly seemed like a simplification, a dream of people who were not well rounded. But this future, it seemed, was inevitable; we must, therefore, bow to these simple solutions, however complex the world.

Especially in the characters of Greg Riker and Nathan Myhrvold, I had an uneasy sense that these new machines were not just being sold as versions of the ordinary tools or labour-saving gadgets to which we had become accustomed by the wealth of the industrial age and, specifically, of the post-war world. They were being sold as the foundations of a new way of living, as a redefining of what it meant to be human, possibly even as tools for changing human nature. The title of Bill Gates's book is strange – not _A_ Road Ahead but _The_ Road Ahead.

In that room where I had been left to collect my thoughts I was waiting not just for Bill but also for the new world to be born. In spite of the slowly forming doubts that would lead to this book, I was still an anxious father who had not been allowed into the delivery room. After half an hour, I was told that Bill would see me.

Today Bill Gates may have been superseded by Apple's Steve Jobs as the hero of the information age, but, even in retirement from the computer business, he remains an emblematic figure of our time. His life falls into two parts: first he was the tough businessman, overthrowing IBM as the primary computer company, then he became the greatest philanthropist the world has ever known, devoting almost $40 billion of his own money to the relief of disease, poverty, hunger and ignorance.

I met him, though I did not realise it at the time, at a critical moment for his business. The Information Superhighway would in fact arrive almost a decade earlier than he expected in the form of the internet for everybody and he was in imminent danger of being outflanked. In that same year, 1994, he sold some of his Microsoft shares to launch the William H. Gates Foundation which was to become the Bill & Melinda Gates Foundation, the world's largest charity. The internet and the later life of Gates were born at the same time.

The future technologies and personalities I had encountered at Microsoft had troubled me but Gates himself was not so disturbing. He seemed to me to be a rounded man, aware that there might be something at least as or more important than technophile immersion. His mind is the modern mind – torn between the values of the spirit and of the ordinary world of getting and spending.

The meeting lasted two hours. We sat opposite each other on low chairs round a small table in his office. His 'workstation' – a word which, like campus, signalled a new type of work – was an L-shaped desk supporting two big screens. Gates looked like the typical nerd, unstylishly dressed in browns and fawns, the haircut of a man who doesn't think about haircuts and the glasses that, as everybody pointed out, always needed cleaning. But when he started talking it became plain he was no nerd. He was an impatient interviewee, grabbing at the core of questions before I had finished asking them. His answers were punctuated with laughter, not because what he was saying was funny but because it was so obvious to him. He had heard every issue and all the opposing arguments before. In a high-pitched and harsh-toned voice, he raced through my questions.

While talking he rocked incessantly. This rocking was not a minor tic, it was a violent motion. He sat forward in his chair, his forearms resting on his thighs and his hands clasped. Pivoting on his buttocks he swung back and forth so hard that the soles of his shoes slapped in a continuous noisy rhythm against the carpet. His pale, staring eyes looked down or directly at me. Occasionally he banged the table with the palm of his hand as if to check that the contents of his head had reached the external world. Sometimes the rocking was not enough to disperse all his impatient energy and he stood up, did a quick five-pace, circular walk and sat down again.

After a while I discovered that the incessant rocking could be stopped by an unexpected question, especially about his personal life. I would say: 'You got married recently ...' or 'I know your mother just died ...' and instantly he froze and the loud voice faded almost to a murmur.

He froze because, at the personal level, he did not want to be known. He argued that his character was unimportant; all the new technology would happen anyway, he was just one more player in the market. He became almost abusive at the idea that anything other than pure reason was involved in business. Twice he slapped me down for using emotional terms.

'No, no, no, there's no need to involve emotion. If you're going to run a business well you'd better not be sitting round here going "Oh yeah, I hate this guy". I mean, come on, it's fun, you should laugh and be very energetic ... You don't get really angry that much ... You don't want to act out of emotion. Fine, maybe I live my life like that, but I don't make business decisions that way.'

So he kept the life veiled behind banalities – 'I do sleep and read and other things' – and I was left with hints and indicators. There were plenty of published anecdotes from the past – portfolios of speeding tickets picked up while driving at 100 mph-plus around Seattle late at night, shadowy girlfriends clinging on to the few hours when he is not in front of a screen, a race with bulldozers hijacked from a building site, a general atmosphere of obsession interrupted by frenzied distraction and of spoilt, unreal boyishness.

He was a divided man. Three-quarters of his mind was work – the rocking – the rest – the immobility – was something unformed, a need to understand and control the world beyond his work. This was two years before Steve Jobs returned to Apple in 1996 to endow computers with aesthetic, quasi-spiritual values. With Jobs, the need was the work. Gates had no such aspiration; his machines were simply tools. But he did want to *know*, to include the immaterial, the ill-defined, in his material, exact world of software code. It was this need in him that started me on this journey.

Perhaps family would answer his need. He had just married Melinda French, a marketing manager at Microsoft. There were rooms in the house he was building for two or three children and a nanny, none of which then existed. Gates was ready for fatherhood; he had written the code, now it just had to be installed. He had even decided his children would not inherit his money – about $6 billion at the time. He planned to give it all away to charity and that, sixteen years later and with three real as opposed to virtual children, is precisely what he did.

'I just don't think it's all that helpful to inherit large wealth,' he said. 'I guess it's a personal philosophy, you have to judge. Everybody spends a lot of time thinking what's good for their kids, I'm sure. And in my case I have decided that having a substantial amount of wealth is more negative than positive.'

But, I pointed out, people change when the children were actually born.

'I don't think I will. But you're right. I accept that that's consistent with what others have told me. But I think there's a little bit of philosophy in this that will not be changed. And I have some friends who feel the same way.' That phrase 'little bit of philosophy' stood out from the usual flow of his conversation. It suggested a still point in the centre of the turning world of the technocrat, something that the eternal novelty of innovation could not alter.

His own upbringing was privileged – a successful lawyer father and a brilliant mother, Mary. Mary in particular was, until her death two months earlier, a constant presence in Bill's life. She died before she could receive the Citizen of the Year award from her local county, honouring her as 'an exceptionally talented, civic-minded citizen'. In 1983 she had become the first woman chairman of the executive board of the charity group United Way of America.

So soon after Mary's death, parenthood was, unsurprisingly, on his mind.

'Parents, yes, absolutely, I think parents are a major influence, maybe mine more than is typical because of my dad's career and the kind of challenges he had and what my mom was doing with non-profit organisations. You can always kind of share those things. They got us to read about the world and talk about it. They treated us as though understanding those things was actually relevant and I went to some pretty good schools and was given a certain sense of self-confidence. There were some things I was good at and I was encouraged to pursue those things. I continued to share with my parents, to spend time with them and value their feedback even as my career got to unexpected dimensions.'

'Even as' suggested that he saw the possibility of a conflict between the values of his parents and the 'unexpected dimensions' of the business he was in, a business built on overthrowing much of the world of the previous generation. Given his claims for the world-transforming powers of the internet, he was being prescient.

Gates, like many successful men, was a gambler.

'Oh yeah,' he laughed, 'I could lose a lot of money. Confidence will do that to you.'

His big bet now was this strange, amorphous entity known as the Information Superhighway or just The Highway as they called it on campus. Bill had an intuition about The Highway: he thought it would be huge, the biggest thing since the telephone, and he had written this book – *The Road Ahead* – to explain why. I asked him why it was called '*The*' rather than *A Road Ahead* or *The Road Ahead in Computers*.

'Intentionally,' he said before I had quite finished framing the question. 'The basic premise of the book is that this advance in communications, where you not only have incredible communications bandwidth but you have these intelligent video devices at the end which are a sort of evolution of the PC although the form factors will be more diverse than we have today. Those will usher in change in a broad set of areas – how we think about shopping, finding information, being educated, or markets or business or living in cities. Do they think of their physical neighbours as their community or use some of these tools to form other kinds of communities with people who have common interests or political views? It's really earth-shaking to have such a revolution in communications.'

This did not, of course, quite answer my question.

His ethical core was partly sustained by religion, but only in the form of a generalised sense of goodness and the right thing to do. He was raised as a Congregationalist, a very American and individualist faith. He said he didn't often go to church but he 'never officially quit'.

'I haven't renounced belief. I think there's some God. In terms of the particular Christian beliefs, I'm not sure about them.

'Yes, I'll bring my children up with a religious background because there's a lot of moral values and there's no substitute for the Church. Even if you don't believe that Genesis is necessarily accurate, there's a lot to be said for the Church. It's a gathering place for people who have a common sense of what's good for the world, what's good for the community. The Church is a place that people come together and talk about things that I believe in.'

Religion is one alternative to the demands of business. Another is art which, like religion, is a potent sustainer of the conviction that there is something more. Gates wanted access to this. It was clear that he thought about art, if only because he knew other people did. He addressed this most lucidly when talking about the effects of the incoming Information Superhighway on cultures. He had been considering the possibility that it would homogenise cultures, turning the world into one big, drab sameness. He spoke enthusiastically and perceptively on the subject.

'That to me is one of the most fascinating questions. The broadcast media absolutely homogenise culture, absolutely. Take the world today, take the distribution of books. Some sell extremely well and then there's quite a tail-off. Take movie rentals. The percentage of movie rentals that are the greatest hits is very high and that's despite the fact that, in terms of any quality measure that you can come up with, there are these classics – movies like *Roman Holiday* – that are better. Somehow people are not aware or they're not interested or they want to see the latest thing that everyone else is seeing even if it's not all that great.

'But the Highway is sort of the ultimate distribution system including looking at reviews – what classic movie should I watch? More niches will grow up – people read Sanskrit, whatever. Or maybe it will make things more centred. People will say: I want to see the popular stuff, I don't care whether it's good or not.

'It comes back to something we don't understand about human interest. It would be better for me to describe the system to a social scientist and then ask him how he thinks this is going to go. We're all human, we're all allowed to speculate. But it's not really my thing. I'm not willing to spend lots of time thinking about this. Just because we're involved in building the system does not mean we know how it will be used.'

I was disappointed that he seemed to back off at the end. It was not likely that a social scientist would be any more convincing on the subject than Gates himself, but the issue was clearly there in his mind, a sense that the intangible world of varied cultures might be changed, even threatened, by the Highway.

He seemed to find art itself a kind of puzzle. When he talked about digitising all the great paintings of the world so that they could be available on The Highway, he did so with the air of a man who was interested in art as a phenomenon, rather than something felt. It was something whose significance in the lives of others he found puzzling. It was a code to be deciphered. And when he spoke directly of the experience of art, he did so in terms of difficulty and the need for explanation. He wanted his technology to smooth the pathway into art. 'The demand that people have is *gated* by the fact that art is hard to approach. Art is full of great stories about new methods and the stories of individual artists. But they are hard to approach. I mean you go to the museum and you see that little label and you wonder why this one is considered better than that or, gee, is that a picture at all?'

He took the ball I threw and ran with it. He did not merely celebrate his technology as world-changing, as so many do, he accepted that how it will change the world was an issue which he was not fully qualified to discuss. There was a certain kind of modesty in all this, an awareness of his limitations that was reflected in his remark as I left his office and he turned back to his screens.

'Back to my virtual world,' he said roguishly.

He was, as I say, a divided man. Perhaps since then philanthropy has filled the one quarter of the world that was not his work, provided the 'more' which he sought. But it was the way he approached art that troubled me. 'Gated by the fact that art is hard to approach' was a strange thing to say. Was it that hard or is approaching art the most natural thing in the world, a meeting of mutually sustaining complexities – the mind and the world? Did Bill see art as a code, a mere complication?

I left the room impressed but uncertain. On the one hand he wrote of *The Road Ahead* as if only the hyper-connected world engineered by Microsoft could possibly be the future; on the other hand he backed off from applying values and judgements to that world, leaving that to the social scientists. At least he knew what he didn't know. But what he did know was plainly going to change my life and the world and perhaps make what he didn't know irrelevant.

Both inspired and disturbed, I began, in my head, to write this book.

2

FATHER'S DAY

The world I had glimpsed in Seattle is the one in which we now live, a world of increasing computing power and connectivity and ever-more intimate machines. But it is not just the world that has been renewed, it is also its inhabitants. Through social networks, mobile communications, cloud computing and many other technologies, humans have begun to adapt themselves to machines in ways that suggest that Microsoft's Greg Riker was right – a new species is, indeed, beginning to emerge. The human self is being transformed.

But there is another side to this picture. While some machines – the gadgets with which we are surrounded – are changing our behaviour, another kind – essentially medical devices – are changing our sense of the human. These are the various types of brain scanners which seem to be closing in on our thought processes. They are, in a way, even more intimate than the familiar little gadgets that steal our information and watch what we buy, for scanners can watch the human brain in action, seeking out traces of the mind within. Specifically, fMRI scans have turned neuroscience into the most urgent and discussed science of our time. This second assault on the self needs to be directly experienced.

* * *

It is 10 p.m. in Sheffield, a city built on nineteenth-century steel and one of the great capitals of the first machine age. A cold, capricious wind is blowing and the almost deserted streets are gleaming from the day's fierce showers. We have walked the half-mile or so from the hospital at a breathless pace, fearing that nothing will be open. It is Thursday, not a propitious day for night life.

Finally, we arrive at a modest row of restaurants. Larry Parsons, the neuroscientist, asks, 'Indian, Chinese or Italian?' There is a pause before I realise that I, the guest, the object of the day's efforts, am the one expected to make the choice.

'Indian.'

We enter The Real Spice.

The menu is too much for me. I pick one dish and then suggest Larry orders assorted starters and Bhavin the rest. I am feeling slightly enervated by my first brain scan. I had been told it would only last forty-five minutes, but, in fact, I was inside the machine for two and a quarter hours.

Larry is our team leader and Bhavin Parekh, apart from being Indian, exudes a certain authority that suggests he will have no trouble with the menu. Tonight he is sceptical of the menu's authenticity, feeling it is far too anglicised, but he does recommend the lentil dishes.

The first half glass of wine, combined with the group's feeling of release from work, at once makes the conversation wide-ranging, fundamental and speculative. There seems to be an unspoken agreement to talk about everything but my brain scan.

The team is enthusiastic. There is the undergraduate Barbara Novakova, who, halfway to the restaurant, announces she must go home. I realise later she may not have been able to afford the meal and I should have pointed out I was paying. There is cognitive scientist Maria Panagiotidi, a Ph.D. student, who says little and who modestly describes herself on Twitter as 'your average geek'. In fact, Larry explains to me, she was the most important player: she constructed my programme of tasks.

Then there is Bhavin. Earlier, we had been looking at a high-resolution anatomical scan – basically a very detailed three-dimensional map – of my brain and he had been pointing out the different parts as Iain Wilkinson, the physicist in charge of the fMRI machine, had made his screen excavate the space between my ears in a series of livid unfoldings, partings before the prow of our virtual neuro-submarine

'That's the brain stem,' said Bhavin, 'the most primitive part. There's the pineal gland.'

I know something about this gland. Descartes, the great philosopher of science in the seventeenth century, felt the dwelling place of the human soul, though perhaps not the soul itself, must be visible in the physical brain. For one reason or another, he chose the pineal gland.

This was, for Bhavin, a baffling confrontation with the bizarre superstitions of the West. He plainly thought I was being serious when I brought it up.

'The soul? No, it regulates things like the onset of puberty.'

'I'm with Descartes,' I said for the sake of mischief.

It had been Iain Wilkinson who, as I had been about to enter the scanner, had taken me aside to make sure I understood the seriousness of what I was about to undertake. This was a legal requirement and, for Iain, a personal matter. While working at the Middlesex Hospital in 1995 on research into the neurological consequences of HIV infection, Iain had volunteered to have his brain scanned as a normal control subject. Control subjects are needed so that abnormalities in patients can be identified.

Iain's scan revealed a meningioma, usually a benign type of cancer that forms on the meninges, a membrane that encloses the brain. It was causing no symptoms at that stage, though it could later have caused blackouts and epileptic fits. The tumour was pressing on his brain and had to be removed. But later it returned, this time with a horrifying symptom – he found he could not speak. Again the tumour had to be removed in a massive operation.

'The main neurosurgeon was stood over my brain for in excess of fourteen hours,' he told me with a touch of pride.

He had good reasons for warning me. They might find something, an 'abnormality', the preferred medical euphemism for a tumour, an agglomeration of gangster cells plotting the obliteration of my world.

'Ethical point,' Iain said once he had ensured I was paying attention.

'Understood.'

In the restaurant I go along with the unspoken agreement not to discuss the scan, partly because I am tired, but also because there is not much to say – the results will not be ready for some weeks and I will be meeting Larry in London to discuss them.

We had talked at length earlier in his office, but this had been a practical discussion about setting up the scanner for my tasks. Larry was seated between his two laptops and one desktop in front of a window. The room was dark and the stormy light outside was lurid and so, throughout our conversation, he was in ever-increasing silhouette, an ominous, outlined ghost waiting to pass judgement on my brain.

But, when not silhouetted and when not matching my details to his computers, he is energetically amiable in skinny jeans, sneakers, pullover and groovy glasses. He has a shaved head and highly expressive hands. He had just had a baby with his French wife and compulsively shows pictures to almost everyone. The laptop on which my soul is to be anatomised has three pictures of his new daughter as its wallpaper.

Larry Parsons is professor of cognitive neuroscience at the University of Sheffield. I had first spoken to him on the phone more than a year earlier when I had been looking into what neuroscience could tell us about human creativity. An American, after qualifying at the University of California at Irvine and San Diego and working as a postdoctoral fellow at the Massachusetts Institute of Technology, he came to England to escape the George W. Bush years and to be closer to Europe. He is one of the big players in the neuroscience game, a highly respected figure.

He is specifically interested in creativity and has conducted some strange, almost comical experiments in an attempt to track the creative pathways of the mind. He has put tango dancers in a PET (positron emission tomography) scanner which also traces blood flow in the brain but does so with the aid of an injection of slightly radioactive water. Of course, they could not actually dance in the confined space, but he did provide a board on which they could do some of the steps. The interest here would be in the dancers' somatosensory or proprioceptive systems, the ways in which the brain is aware of the position of your body. If you move your arm in the dark, for example, you can still sense where it is so there really is a sixth sense. This was to prove an important – in fact, startling – aspect of my results.

The pop star Jarvis Cocker was also stuck inside his machine as a subject for studying musical improvisation. He had come up with plenty of information about what parts of the brain light up when people are being creative. But, Larry admitted, there's not much we can do with this. Neuroscience lacks a big theory, a way of making sense of the information that pours from the scans.

'It may be three hundred years before we can do things like enhance our creativity,' he told me with a gloomy chuckle. 'Some say in twenty years we can make you smarter but I'm a pessimist.'

This was the third reason Larry was the man I wanted to scan my brain – his sceptical and realistic approach to his subject.

'There is this over-complicated thing [the brain] we barely understand,' he said during that first phone call, 'because we're only at the beginning, we're still looking at the circuit diagrams.'

There is a huge gap between the blood flow in the brain and the phenomenon of consciousness; the hard problem remains as hard as ever.

Our job in the afternoon before the scan was to write a Power-Point presentation of all my tasks. This had to be precisely timed as I would see the presentation while I was in the machine and each task had to last a fixed period of time. The PowerPoint had to be synchronised, second by second, with the machine which would be set up to run precisely as long as each phase of the presentation. It was all very complicated and I became anxious because I could not possibly remember everything Larry had said. But, he assured me, it would be okay, he would be in the room with me. He would tap my foot at critical moments to indicate I should move on to the next stage. The scan, he said, would last forty-five minutes.

Then he told me to go to Starbucks across the road. He needed to set up the PowerPoint on his computer and I could not be allowed to see the questions and tasks I was to be set. It would spoil the results if I had worked out my responses in advance.

The Starbucks was housed, improbably, in one of those solemn northern houses built of forbidding dark stone. Inside, it was the usual globalised, cosy scene of excess – strange coffee variations, some topped with coils of cream, the drunkenly puffed-up cakes they call muffins and the indigestible slabs of spongy chocolate they call brownies.

Amid a crowd of socialising and studying students, I spent an hour and a half in the café reading Semir Zeki's *Splendors and Miseries of the Brain*. I had seen Semir, a neurobiologist at University College London, the day before and something he said about Larry and that Larry had just said about him had suggested to me an element of mutual disapproval. But, perhaps, only an element and, anyway, self-respecting academics seldom approve of others in their field.

They do have much in common. Semir is one of the old masters in the field of neuroscience. He is also the man who invented and defined the field now known as neuroaesthetics, the attempt to discover the roots of creativity, art and beauty in the physical fabric and workings of the brain. Larry scans brains to see what happens during the creative process; Semir studies great art to discover what it can tell us about how the brain works and then uses scans to test his insights. They are approaching the same thing from different angles.

Finally, Larry arrived, grabbed an espresso and said we had to wait for Bhavin, Maria and Barbara. More waiting, but, happily, they arrived almost at once and we plunged out on to the now darkened street and then into the labyrinthine ways of the Royal Hallamshire Hospital.

There are two scanners here; one is shared with the university. Yet this was still a hospital, not an academic environment, and I was the patient. Racing to keep up with Larry's lengthy stride, we passed people who were actually sick. I felt like an imposter. But Larry sailed on, his students trotting behind. He had to stop and engage in banter with various staff, showing them pictures of his baby on his phone.

Then there was a door covered in warning signs leading into a room full of computers from which led another door covered in more warning signs. Entrants must not have a pacemaker, artificial hips, wear watches or carry credit cards. I had to fill in forms about my health and to be checked for metal. The studs on my jeans were all right because they were firmly fixed and perhaps Levis don't use ferrous metals. Who knows? The forms asked me what medication I was on, something about my medical history and contact details for my doctor. It was routine stuff but, somehow, it seemed more difficult and important than usual.

Time passed with much extravagant camaraderie. I was relaxed, feeling like a celebrity, pampered by his posse. But then the inner door was opened and I shuddered slightly as the MRI machine – a giant, fat ring surrounding a narrow tunnel – appeared. It takes so much effort, so much ingenuity, so much heavy machinery, I thought, just to glance inside the feeble electrochemical system of our ordinary little brains. But the shudder was also to do with the fact that, as one of the tests, I had decided to discuss my father's depression and death, subjects I have been unable, until very recently, to discuss with anyone. I was dreading this.

The lights in there were dim and the whole thing had a religious look – the altar of some initiation ceremony, perhaps; or sc-fi – the entrance to a worm hole that exists in some distant galaxy. It also seemed somehow autonomous, as if, when not scanning, it would have other, perhaps more important, things to do.

The first nuclear magnetic resonance demonstration took place in 1946 in the United States. In 1952 Felix Bloch and Edward Purcell won the Nobel Prize for their discovery of the basis of what would become NMR spectroscopy which was used to analyse chemical compounds. Like MRI, this works because, in a magnetic field, the nuclei of some atoms absorb energy and then, in flipping back to their original state, emit this energy. This emission is what was being detected. In 1973 the potential medical application of this was made clear in a paper by Paul Lauterbur, a professor of chemistry. He called it it zeugmatography. He showed that magnetic resonance could be used to locate two test tubes of water. This spatial element was the basis of an imaging technique and it transformed NMR into MRI. The machines were developed in the 1980s and became, in the 1990s, commonplace, the primary tools of modern medicine. Place a patient inside an enormously powerful magnet and the emissions as the nuclei flip produce a phenomenally detailed picture, far beyond the fuzzy shadows generated by the X-ray machine and, unlike them, including soft tissue.

The machine I was about to enter was a 3 Tesla (3T) which is twice as powerful as most hospital scanners. These 3T scanners came on the market seven years ago and this one was one of the first in the UK. To give some idea of its power, the earth's magnetic field in which we spend our time is 0.00005T. Iain told me that 7T scanners will soon be available, plunging patients into an unimaginable magnetic maelstrom.

Higher power means better images. The scanner was going to watch the behaviour of my protons, positively charged particles in the nuclei of atoms, specifically the protons in hydrogen atoms.

My hydrogen nuclei could either line up in the direction of the field if they were in a lower energy state or against it if their energy state was higher. At normal body temperature slightly more line up with the field. The stronger the magnet, the more line up. A radio signal is then sent in and the lower energy, lined-up nuclei flip to the high energy state. When the signal is turned off, they flip back, emitting energy. It is these emissions that make the image.

It seemed impossible that such powerful magnetic fields should not cause permanent damage. I had put this to Larry.

'We're not magnetically sensitive like pigeons,' he had said. Pigeons are thought to navigate using the earth's magnetic field. I wondered if he had ever put a pigeon in there. The bird might never again find its way home.

In fact, the most likely threat to scannees is ferrous metal objects accidentally left near the machine. These would fly like bullets towards the scanner, a phenomenon that has caused injuries and occasional deaths. In 2001 Michael Colombini, aged six, was killed in an MRI machine when an oxygen tank flew out of the hands of a technician and hit his head. At Sheffield, however, they seemed pretty cautious.

I am to have an fMRI scan. This detects changes in blood flow in the brain. Blood flow was known to be linked to neural activity before scanners came along. The parts of your brain in use need more oxygen and, therefore, more freshly oxygenated blood. This produces hot spots on the image which can be correlated with whatever the subject happens to be thinking at the time. This means the brain can be mapped safely and precisely, simply by matching what you are thinking or doing to the little starbursts and streaks that appear in the scanner's images. The fMRI system is the foundation of the great edifice of neuroscience that is now being constructed around us.

I was in the outer room for well over an hour as Larry, Iain and the students worked to sync the PowerPoint and the MRI.

Periodically I heard a screeching-grinding sound from the machine.

Finally, Iain took me in, issued his solemn warning about the possibility of finding something seriously wrong with me and then taped a cod liver oil capsule to my left forehead. This shows up well on the scans and it makes it easier to tell left from right when they are being studied.

'Then you can eat it,' said Iain, 'and it will do your joints good.' I thought that was an old wives' tale, but maybe . . .

Iain gave me ear plugs and then a large pair of headphones. I began to worry that I would not be able to hear Larry. I lay down on the scanner's sliding table. A big wedge was placed under my knees for comfort and I laid my head in the head-shaped hollow. Iain Velcroed a tape round my forehead to keep my brain still and then encased my skull in a strange helmet-shaped contraption. My nose just touched its surface. There was a mirror over my head which would allow me to read the screen that hung a few feet beyond my feet. Iain laid a panic button by my right hand but then pleaded with me not to press it as everything would then have to go back to the beginning. He then slid me back towards the machine, stopping just as my head was entering the tube. He told me to close my eyes while a laser established the exact position of my nose. Then I was slid fully in.

Larry took pictures of me with his iPhone just before I finally entered the machine. He sent them to me later and, for the first time, I became fully aware of the strangeness of the scene. In the best picture I was lying on the slider with the tunnel behind my head. Around my head is the helmet contraption with the mirror held over my forehead by what looks like a robot arm. Just above my feet is the big screen on which my tasks and tests will be displayed. Behind the screen is the window of the outer room where Larry, Iain and the students would struggle with the complications of the scanner controls and the timing of the Power-Point.

From inside the tunnel I had two views of the outside world. One was provided by the mirror, the other was only possible if I peered down the length of my body. This allowed me to look over and between my raised knees into the room which, for me, had been reduced to a gloomy semi-circle in which Larry's head periodically – and comically – appeared to check on me or give me instructions. In the event, I could hear him quite well.

The space was horribly small. The diameter of the tube was a mere sixty centimetres, about the size of a largish bucket. On first entering, I was hit by a wave of panic. I am not normally claustrophobic but this was very extreme confinement. Also, my face had begun to itch all over. It took me half an hour to discover I was actually able to scratch it. I had been told to keep still so often that I was afraid to move anything.

Strangely, my panic was controlled by a sudden thought. This was, I realised, an approximation of the conditions NASA's first astronauts had to endure in their Mercury capsules. With near certainty, I knew I could get out, but John Glenn, Alan Shepard, Wally Schirra and Gus Grissom would have had no such consolation. Three things could have happened to them with almost equal probability – they could land safely, burn up or drift endlessly until they suffocated.

Then Iain fired up the machine for some sort of test run. At rest there had been a distant pumping sound like an old steam engine waiting in a station. But now there were some less benign, though equally industrial age noises. There was an awful scraping – like a crowbar gouging a groove in a girder – followed by thuds which were deafening even to my protected ears. A couple were so close they made my head vibrate. A demented troll seemed to be hammering the giant ring around my head. Then there was screeching I had heard before but much louder.

This was, I assumed, the actual scanning. How could anybody

think clearly in all this noise? Maintaining concentration through my tests would be difficult.

The noise happens because the machine acts like an enormously powerful loudspeaker. Iain said he has heard Bach played using the scanner as an instrument by employing its controls as a kind of keyboard. The 'gradient coils' in the machine have to be turned on and off during scanning to make the images. The coils are encased in epoxy resin concrete but the changes of current in a magnetic field still cause them to move, which is what causes the noise. It has so far proved impossible to design the noise out of scanners.

The content of the tests had been discussed in advance. There were five phases: humour, poetry, lying, emotional and creative. They had been tailored to me as a writer and on the basis of emails Larry and I had exchanged over the previous few months. I was particularly keen on the creative tests. The more I had been thinking about the scientific attempts to solve the hard problem of consciousness, the more I had become convinced that creativity, art and beauty formed the core issue. These, surely, were the values and ideas that clung naturally to the word 'soul', and the very existence of that word is clear evidence of our ancient intuition that the mind is not simply the brain, that there is, indeed, something more. But that raises the question: what is it we are watching on the MRI scans? Traces of the mind?

My programme was, I later learned, one of the most complicated they had ever done. As a result, Larry had been over-optimistic about how long I would be in the machine.

The humour test was easy but slightly strange. I was shown a series of cartoons. These were the standard little dramas – husbands and wives, men in offices, women in shops. They were funny enough. But then, after a short pause in which I was to think of nothing in particular, I was shown a series of cartoons that were deliberately not funny. This was odd. The cartoon convention of a picture with speech bubbles or a caption so strongly signifies a joke that I found myself thinking I had missed something, that I didn't get it.

This combination of the test itself combined with a neutral, control version was the same each time. In order to tease out the brain activity that results specifically from the test, it is necessary to check the brain activity when that is not part of the test. So the unfunny cartoons show what my brain is doing when I look at a cartoon-like picture without being amused. The difference between this and my response to the funny cartoons was what was being measured. Similarly, in the ensuing truth-lie test, I was asked to give a true account of what I did the day before and on my daughter's wedding, which had happened six months earlier. In each case, after a pause, I was asked to give completely false accounts of the same things. This was surprisingly difficult. I tried to keep the lies plausible. Something went wrong with the scanner, however, and I had to do the whole thing again. This time I veered off into fantasy – cows on railway tracks, mass punch-ups breaking out at the reception. Fantasy proved significantly easier than plausible lies, a general truth about the fictional mode.

The final lie test was the most demanding. I had to answer sixty questions, fifteen slides with four on each, first truthfully and then falsely. The questions were banal – how much did I weigh, what colour were my wife's eyes? – but this made it even more difficult to lie. The familiarity of the answers made them so obvious that coming up with anything else was like lying to myself – I had to tell an outrageous lie to someone who knew exactly how outrageous it was.

For the poetry section, I had chosen two poems that I knew and loved and these were to be alternated with more or less nonsense poems, chosen to provide completely opposing responses. My two poems were Wallace Stevens' 'The Snow Man', an appropriate meditation on the nature of perception, and Emily Dickinson's 'The Brain is wider than the Sky', which speaks for itself.

The emotional tests were the most disturbing. Here I had to recount two memories that were, for me, powerful. The controls were two relatively neutral memories from the same period. My neutrals were tales of playing cricket, a game at which I am neither bad nor good enough to provoke strong personal feelings. But, in contrast with these, I had decided to tell two stories about my father – about his depression and death. I told these, unscripted, inside the machine. I was allowed a minute for each. What follows is a more polished and longer version of what I was trying to recapture.

My father, Cyril John Snowdon Appleyard, died when I was thirteen. He had been senior consultant design engineer at ICI and he came from a family of distinguished scientists. He was, to me, the role model of high intelligence and wit. He had been everything to me and I had no idea his life was in danger, though I did know his asthma had grown progressively worse. I, too, had asthma. This was before the lung-opening sprays that were to transform the lives of sufferers and, throughout my childhood, my sleep had been interrupted by attacks that a tablet, Franol, took twenty minutes to quell. Now children can take Ventolin and usually find relief at once.

But my first memory was not of his death. It happened a year or two earlier. It remains for me an intolerable vision of utter bleakness and despair, one that, until quite recently, I have been entirely unable to recount, even to myself.

He suffered from depression. This had, unknown to me, worsened. He had been prescribed electroconvulsive therapy (ECT) in which the brain is given an electric shock which, it is hoped, will cure the deeply depressed. It sounds brutal, a treatment from the industrial age, and it is. ECT remains in use, though there are controversies about its efficacy.

I was about twelve. We had always been a family without a car but my brother, Richard, had just got his driving licence and he had a red Morris 1100 with a light grey interior. I knew that we were going to pick my father up from hospital and I think I knew an electric shock was involved. I was in the front seat next to Richard and my mother was in the back. We stopped outside the hospital and, after some arrangement which I do not remember, my father was seated in the back seat.

He said nothing; none of us said anything. I looked back at him. He looked smaller, somehow, and his absent expression made him a stranger. In shock, I felt a cold prickling in my skin and my imagination withdrew from the scene in a desperate attempt to protect my feelings. He looked defeated, forlorn, this man who I thought knew everything.

Sitting in the car, I had a vision of what they had done to my father. Men in white coats had sent a jagged blue flash cutting through everything I most loved in the world: his brilliance, his science, his engineering, his sentimental love of Kipling and odd quotations from *The Rubaiyat of Omar Khayyam*. There was one couplet he was always quoting – and which now still runs through my head – which I once took to be Kipling but was, I now discover thanks to the internet, from Laurence Hope's *The Garden of Kama*.

I am weary unto death,
Oh my rose with jasmin breath.

The memory of his death – my second story in the machine – is equally vivid, but, curiously, less harrowing. My aunt, uncle and cousin had spent the evening with us and left. I went to bed. I woke to see a pale oval hovering in the half-light of the January morning. It was my cousin's face; he was lying on the other bed, staring at me.

'I thought you went home.'

'We came back.'

We came back! Impossible. I knew, something, everything, had gone wrong but I said nothing. My cousin got me into my dressing gown and ushered me down the hall. I made a move towards my parents' bedroom and I felt the sudden pressure of his shoulder, pushing me away. Something was certainly wrong. In the living room my mother was sitting on a sofa, my aunt was there. I sat down next to my mother and she put her arm round me. This also felt wrong; she was seldom spontaneously affectionate. There was a pause. I had to say something. I recalled the exact words as I was lying in the machine

'What happened last night?'

My mother called my aunt over.

'He wants to know what happened last night.'

My aunt knelt before me and put her hands on my knees.

'Bryan, last night your father died.'

I rose and started to walk across the room; my aunt stopped me and hugged me. I wept copiously. But *the tears were fakes, every one.* My entire being had skipped out of my body. I had a distinct sense of jumping to one side in a way that was invisible to everybody in the room. I had immediately decided I was not going to be the person whose father had died. It had happened, *but not to me.* I was watching my grief from a great distance, cold and unconcerned. I was not to find my way back into my body for many years.

And so I told my stories amid the industrial age screechings of a device that was associated in my mind with the machine with which they had electrocuted my father. The stories were about my fear of medicine – especially when machines were involved – and of physical assaults on the brain. I had become, in my imagination, my father undergoing ECT, a man lying down helpless beneath the cold workings of a machine that was said to cure.

Because of the power of the emotions involved, I did not just rest in between narratives during this phase. I also had to count backwards from one hundred for fifteen seconds. This was thought to be sufficiently difficult to neutralise my feelings for the next task. But, strangely, my feelings seemed muted. Normally I cannot tell those stories without weeping. This time, I did. The noise intruded and the countdowns did, indeed, distract me. I wondered if fMRI scans could even begin to capture such things.

Then the creative test. My neutral creative task was a boring ponder about how to pack the car to drive to our house in Norfolk. My real task was the invention of a short story from scratch. My previous day's conversation with Zeki had strayed into the area of courtly love – the medieval convention of the chaste romance. This came back to me and I outlined a story about a modern man who determines to pursue a courtly affair. I got as far as giving him the idea and sending him to the London Library to research the subject but I ran out of time before I could introduce the lady he had decided to worship.

Finally, there was the long anatomical scan. This was simply a three-dimensional picture of the whole brain. I had to keep perfectly still while the machine emitted a new dull bass grinding. I even felt elation at this motionless inactivity.

At last, I was sliding out and the students were gathered at the door. I asked Iain how long I had been in. He said, 'Oh, about an hour and a half.' I asked Bhavin exactly how long I had been in. He studied the clock earnestly.

'Two and a quarter hours,' he said firmly, 'it is the longest we have ever had.'

If he had said half an hour I would have believed him. I had lost all sense of time. But he was right, the clock on the wall said 9.30.

I watched as Iain studied the anatomical scan on this screen and listened as Bhavin told me what I was seeing. After my mischief about the pineal gland, I confused him further by saying the pictures were disappointing, as X-rays images were always disappointing. It is their impersonality, their lack of character. Inside, we look like everybody else, like the skeletons and lurid diagrams of muscles, nerves and organs we saw at school. Strip us of our skin, our features, and at once we become a crowd, weaving its way to oblivion.

And so we came to The Real Spice. We are the last to leave, the poor, resigned waiters are muttering to each other in a corner. This is a student area so they have probably seen many such long, drunken discussions about impractical matters like meaning and purpose. I have still not spoken to Larry in depth about neuroscience, scans and what they mean. I decide to leave it until we meet again in a week or two to discuss the results.

Later, trying to sleep, I remember the Emily Dickinson poem again. The second verse repeats and twists further the idea of the first.

The Brain is deeper than the sea –
For – hold them – Blue to Blue –
The one the other will absorb –
As Sponges – Buckets – do –

* * *

A few weeks later, Larry and I meet to discuss the results of my scan in the café in the gardens of Russell Square in the centre of London's university district. A recent heavy fall has left the square snowbound and the café is draughty. We talk for two and a half hours, slightly longer than my scan took.

Patiently, Larry tells me about the structure of the brain, its gyri – hills or ridges on the cerebral cortex – and its sulci – crevices or valleys, as well as the less poetic numerical system for identifying different regions. It is a geography lesson; I am being taken through a dangerous and uncharted land of hills and valleys. The brain is described like this – nobody talks of hills or valleys elsewhere in the body – because its shape is irregular and strange and the functions of each region are never quite clear. We remain explorers in this land.

The brain is also mysteriously doubled; there are two almost identical halves. Some believe this represents a fundamental division between two sides of the human character – emotional and rational – but Larry is cautious and resorts to 'we' to denote the neuroscientific community. 'At a first approximation,' he says, 'everything talks to everything else. We do think that there are special localised functions, but to believe a whole hemisphere has a style of doing something is a very strong claim. We don't believe that for the most part.'

If brain structure is a geography lesson, then going through my results combines computer science, mathematics, psychology and, finally, neuroscience. Larry opens his laptop and starts showing me images of my brain with apparently random areas luridly coloured as if by an inattentive child. He explains that the colouring shows which areas were especially active. But it is not that simple. The results of control tasks – the boring stories, the unfunny jokes, the truth telling – have been subtracted from the results of the interesting ones, leaving a picture of the distinctive activities associated with interesting stories, funny jokes and lies.

As he speaks, I become aware of an oddity in the way he describes what happened inside my head. 'Here,' he says, 'you are accessing this emotional area.' This suggests that there is somebody – 'you' – wandering around my brain, finding the right area to activate. But where is this 'you' if not in the brain? All the areas of the brain do something, but none appear to be the mind that is thinking. But we say I access this or that area. Our language is constructed to define the human self as an invisible ghost in the machine of the brain, a soul.

The actual results are, at first, predictable. In response to the cartoons I 'show more or less the same patterns of activity' that have been identified in other scientific papers.

'The standard theory of humour,' says Larry, 'is something like there's some set-up which creates some kind of tension in thematic ambiguity then there's a punch line which disambiguates the tension in a way that is unexpected.

'Humour is hardest to transmit across cultures. There are embedded empathic assumptions that are social in nature. It's the last aspect of a new culture in which you become fluent. It involves emotion and cognition.'

He then makes a point which casts doubt on the meaning of brain scanning.

'There's nothing exact or precise. It wasn't as if we found a joke that only you would get. It wouldn't show some activity that was specific to you. Your experience of humour would be entirely different from mine and yet you might still have the same brain states. All we are getting is blood flow. So fears about mind reading are ludicrous.'

This is another way of looking at the grammatical problem. Again a ghost appears in the machine. If Larry and I have the same brain states and yet different experiences, where is the person who is having the experience if he is not in the brain state? Larry would not be his brain state and I would not be mine. You can't read my mind even if you can read my brain.

The poetry test results are a surprise. I actually seemed to be more emotional about the scrambled, nonsense poems than I did about the Wallace Stevens and the Emily Dickinson. Larry consoles me by saying that it may be something to do with the way the results were derived from the scans, but then adds, 'The poem test didn't quite give us that extra deep element we were hoping for.'

My results in the lying tests are similarly puzzling. There were not as many active areas as there should have been, either, suggests Larry, because I am a really bad liar or I am so good I couldn't tell the difference between lies and truth. I avoid responding to that.

The emotional memories of my father's electroconvulsive therapy and his death also produce surprising results. There was much less of an activation of the emotion areas than would have been expected. But there is one oddity. During my account of his death, my proprioceptive system was activated.

Proprioception is all to do with body position and movement and is, in fact, a kind of sixth sense that allows us, for example, to know where our arms are in the dark. You do not have to see, feel, smell, touch or taste your arms to locate them: you just know where they are. The finding suggested some strong movement component in the memory, which puzzled Larry. It did not puzzle me. The scan, I decided, had picked up my feeling of jumping out of my body at the moment I was told of his death. Although, of course, I did not actually jump, the feeling of moving quickly to avoid this calamity was overwhelming. Larry purses his lips and does not argue.

Composing the short story produced the most brain activity. Using several of his coloured pictures, Larry runs through lists of areas activated – working memory, meanings semantics, planning sequences, language context, resolution of conflicts, looking for error inconsistencies, nested meaning, learning, intrinsic memory, emotional areas, the hippocampus – so you were drawing on memories, somatosensory indicating movement in the story. Creating something new evidently engages the entire brain.

The leap out of the body and the fireworks produced by the short story composition were, for me, intriguing results. The rest were either perplexing – the poetry reading – or routine – the cartoons. But such disordered results are, in a way, the point. This is plainly a science in its infancy. As Larry had said, 'we're only at the beginning, we're still looking at the circuit diagrams'.

Some time later, I also spoke to the neuroscientist Colin Blakemore. He compared our current scanners to the crude telescopes that Galileo used. He saw things we now take for granted – the moons of Jupiter, the mountains on our moon – but now our telescopes can see deep into space and resolve the fine structure of the universe. The pictures we see of our brains are blurred and low-resolution. In future decades and centuries they will spring into sharp focus and high resolution, then what will we find? My rather strange results could be no more than evidence of the machine's imperfections, smudges on the lens of the telescope. They did not mean my mind and soul were necessarily beyond the reach of the machine.

And so, in a sense, the way neuroscientists talk of emotion, memory, language or learning areas is misleading. They are not necessarily wrong to do so but we are wrong to assume that these words are anything more than a very basic map of this rough terrain. And it is one thing to look at a road on a map, quite another to walk down it.

Then there is the strange problem of human consciousness. This is famously difficult to define and, indeed, for many years scientists and philosophers seemed to give up. The 'hard problem', of how matter becomes mind cannot really be solved unless we can say exactly what we mean by consciousness. It is plainly present throughout the living world in the form of a hierarchy. There are grounds for arguing that a bacterium has a form of consciousness, but it is obviously less conscious than a gorilla. Human consciousness, however, seems to be entirely different. Gorillas spend their time in troops in the jungle; humans build cities, write symphonies and ponder the meaning of it all.

We seem to be self-conscious, we can think about ourselves and we can think about our thinking. It is an ability probably closely associated, either as cause or effect, with language. This is barely a definition; it is more of a statement of what is to be defined.

But the point about self-consciousness is that it changes everything. When Larry says 'you are accessing' a brain area, he has to use language in such a way that there seems to be some ghostly me that is not made of brain matter. So even though I am sure he thinks the self is right there in the brain, the fact of self-consciousness means he cannot avoid speaking as if it is not.

Many philosophers have got round this by saying the self is a delusion. The brain uses the consoling idea that we are the same person from moment to moment to make us more efficient predators and breeders. But this runs into the same problem, for, if the self is a delusion, *who* is being deluded?

So what were those blotches and streaks on the pictures of my brain? Were they the cause of consciousness or its effect? Or did they just indicate that *something* was happening in the brain without giving any clear indication of what? Either one day science will answer these questions or they are unanswerable and the one thing the self-conscious mind cannot know is itself.

The human brain, people say, is the most complex thing in the universe. But its complexity is nothing next to that of the human mind.

3

HENRY

Back in the 1980s, Larry Parsons used to know a man called HM. Larry was a postdoctoral fellow at the Massachusetts Institute of Technology at the time and, every six months or so, HM would arrive and stay at the intsitute. Scientists would then subject him to tests. HM seemed immune to boredom or irritation; he cooperated with irrepressible good nature. Larry remembers him well.

'He was a sweet man, he was so good. He was in fine health and he was only in the hospital because he couldn't walk around. He wouldn't know what he was doing.

'He sort of knew he had a problem . . . He had a sense of humour, he could make jokes but he knew he didn't quite remember things from moment to moment. He would do the same crossword every day.

'I would come into the room and he didn't know if he had ever seen me before. So he would just say "Nice to see you". He was in this grey zone of complete doubt but he would just cover it up.'

In medical and scientific circles the patient was known only as HM to protect his privacy. Now he has no privacy to protect and his full name, Henry Gustav Molaison, is widely known. He died at the age of eighty-two in December 2008.

At the moment of his death, a carefully rehearsed routine swung into action. Jacopo Annese, director of the Brain Observatory at the University of California at San Diego, was informed and he boarded a plane to fly 2,600 miles eastward to oversee the removal and preservation of Henry's brain. It then had to be chemically fixed and hardened over a period of weeks. Finally, the brain was packed in a fluid and cotton wool-filled box ready for transportation to San Diego. There were airport security issues with a box full of so much fluid but the brain was eventually allowed to board. It took the window seat, Annese took the aisle.

Subsequently the brain was sliced into 2,401 sections, the whole procedure being watched by 400,000 online viewers. All the slices are now available for study on the internet. Henry's brain has become the symbol of our contemporary search for the material basis of the human mind.

From childhood, Henry had suffered from epilepsy, possibly caused by a cycling accident at the age of nine. His seizures steadily worsened until, by 1953, he was utterly incapacitated, at which point he was referred to a neurosurgeon, William Beecher Scoville, at Hartford Hospital in Connecticut. There were high hopes for direct medical interventions in the brain at the time. Lobotomies – cutting the connections to the prefrontal cortex – were widely performed. The administration of ECT, meanwhile, was commonplace; my father's case was not unusual. In Montreal, the neurosurgeon Wilder Penfield had been reporting success in treating epilepsy with a procedure that involved destroying brain cells at the point of origin of the seizures. It seemed like a simple solution.

After consulting with Penfield, Scoville decided to suck out brain matter from the area in which Henry's seizures originated – around the hippocampus, a worm or sea-horse shaped component deep in the brain. The procedure worked well as far as the epilepsy was concerned, Henry's fits subsided. The side effects, however, were catastrophic.

Only parts of the hippocampus were removed, but, after the operation, what remained was disconnected from the cortex. Furthermore, surrounding structures were removed. We now know the function of the tissue removed by Scoville is to lay down long-term memories; it does not store them, but sends them to other areas of the brain. Henry became entirely unable to form memories. Doctors, scientists and nurses who visited him daily went unrecognised, though he did claim to have met one scientist in high school. His memories prior to the operation were partly intact, so he had simply pushed her back in time to the first twenty-seven years of his life. He also retained good semantic memory and was able to complete crosswords. Apart from that, Henry lived in a perpetually novel and inexplicable present in which each new event persisted in his mind for less than twenty seconds.

The cataclysm which befell him is made more poignant by the man's character, both before and after the operation. Prior to surgery he was said to be an intelligent, amiable man with a liking for ice skating and mystery shows on the radio – he was said to be good at spotting the villain before the detective. After surgery, the essential serenity of his character remained intact.

'He was a very gracious man, very patient, always willing to try these tasks I would give him,' said Dr Brenda Milner, a cognitive neuroscientist who worked with Wilder Penfield, 'and yet every time I walked in the room, it was like we'd never met.'

Bizarrely, after the operation, his IQ increased from about 110 to 120. This is thought to be because of the reduction in the amount of anti-epilepsy medication he required.

While still living at home with his parents, he could get through the day and perform ordinary tasks – shopping, mowing the lawn, raking leaves, preparing lunch, making his bed and watching TV – on the basis of the memories he had formed before the operation. Even when in a nursing home and being studied intensively by scientists at MIT, he retained a touchingly modest sensitivity. Milner has spoken of an occasion when a researcher said, in Henry's presence, what an interesting patient he was.

'HM was standing right there,' said Milner, 'and he kind of

coloured – blushed, you know – and mumbled how he didn't think he was that interesting, and moved away.'

'He was like a family member,' said Dr Suzanne Corkin, who was in charge of the HM research at MIT; 'you'd think it would be impossible to have a relationship with someone who didn't recognise you, but I did.'

There was, however, also confusion and unhappiness. He never knew how old he was but usually guessed about thirty. Every time he saw an older man in the mirror he was surprised.

There was something both charmingly ordinary and yet painfully exotic about Henry Molaison. His life continued in more or less familiar ways and yet he remembered nothing after the operation. Every moment was lost. The thought of the present continuously falling into this dark void, of the pre-mortem past being no more than a presentiment of the post-mortem future, induces a sickening vertigo in anybody who hears Henry's story. His plight strikes at our deepest sense of who we are. Memory assures us of our continuity through time; without it we seem to become nothing but fragments.

'What HM lost,' said neuroscientist Thomas Carew, 'we now know was a critical part of his identity.'

The idea that a simple change to the physical structure of the brain should, on the one hand, take away so much and yet, on the other hand, leave so much intact is bewildering.

Memory is so close to what we are that the loss of it grips the imagination. The hero of Christopher Nolan's film *Memento* (2000) is in a similar predicament to Henry's except that he is in pursuit of his wife's murderer, that event being the last thing he remembers. As a result, he can only make desperate attempts to link one moment to the next by leaving himself notes.

But, if too little memory is terrible, too much can be equally appalling. Ireneo Funes, the central character of Jorge Luis Borges' story 'Funes, the Memorious', remembers everything, a condition that induces terror in the narrator, a terror of creating yet more memories to plague Funes:

> Then it was that I saw the face of the voice which had spoken all through the night. Ireneo was nineteen years old; he had been born in 1868; he seemed as monumental as bronze, more ancient than Egypt, anterior to the prophecies and the pyramids. It occurred to me that each one of my words (each one of my gestures) would live on in his implacable memory; I was benumbed by the fear of multiplying superfluous gestures.

Henry's case had further depths. Repeated testing of the man – he was relentlessly patient as he had no awareness of the tedium of repetitive tasks, forgetting each one within seconds of its completion – revealed that he could form certain kinds of memories. He could acquire new motor skills. A drawing test that he found difficult at first became easy after a few days because, though he knew nothing of this, he retained the capacity to store physical movements. He could also retain spatial memories – he was able to construct a mental map of the house where he lived with his parents until he was moved into a care home in 1980. Such memories were formed, apparently, on the basis of daily movements and may therefore have been related to his ability to retain motor skills.

Henry was a perfect research subject and not only because of his patience but also because of the purity of his amnesia which co-existed with a perfectly intact intellect. Recordings of his voice reveal a supremely amiable and peaceful individual. He had no psychiatric symptoms and was eager to help the scientists who queued to study him. For a few seconds at a time, he was proud to be the subject of their interest. He hoped he could be of help to others. His graciousness may have had something to do with the date of his operation. As his memories stopped at 1953, when

Eisenhower succeeded Truman as president and the first fruits of the post-war economic boom had begun to rain down, he was, in his strange way, a living memorial to what is still seen as a 'kinder, gentler America', an always newly arrived time traveller from the past.

Henry may be the most important patient in neuroscientific history. The centrality of the hippocampus and surrounding areas in memory formation and the discovery that motor skills were retained in a different way from 'declarative memory' – the recollection of new experiences, names and faces – were momentous breakthroughs. Direct study of his physical brain is now likely further to revolutionise neuroscience.

The most obvious but also the most important point about his condition was that it was caused by surgery which had resulted in very clearly defined damage to his brain. The destruction of his ability to form memories could, therefore, be shown to have a direct link to an area in the brain.

'Prior to his surgery,' said Corkin, 'the clinical literature contained hints that this area played a role in long-term memory. His case, however, showed definitively that the hippocampus and neighbouring cortex are critical for the establishment of long-term declarative memory.'

Henry is thus the scientifically triumphant climax of a long history of bizarre and often politically motivated attempts to link the physical brain to the mind – in other words, to solve the hard problem of consciousness and finally locate our minds in our physical brains. Some of the earlier attempts betray a degree of desperation about the project. Perhaps the most misguided was the effort to prove a link between brain weight and intelligence.

The brain of the Russian novelist Ivan Turgenev weighed just over two kilograms, that of the American poet Walt Whitman 1.28 kilos. In fact, the latter figure is an estimate, as Dr Anthony Spizka explained to the New York Times in September 1912:

'The weight is sheer guesswork as it stands. The brain had been preserved, but some careless attendant in the laboratory let the jar fall to the ground. It is not stated whether the brain was totally destroyed by the fall, but it is a great pity that not even the tiniest fragments were rescued.'

Spitzka ran the American Anthropometric Association in Philadelphia which, post-mortem, collected the brains of clever, successful people for study. Anthropometrics is the study of the size and proportions of the human body. Spitzka was convinced that brain weight and brain tissue of 'superior organisation' were clear indicators of superiority in life. As evidence, it was pointed out that the brain weight of an 'average bushman' – say, a Zulu – was said to be as low as one kilogram. The very light brain – 1.16 kilos – of the radical, heroic and world-famous French statesman Léon Gambetta was plainly anomalous.

The Soviets agreed with the principle that human excellence ought to be visible in the human brain. Communism is, after all, a strictly material creed so it seemed natural to suppose that the brain of its greatest hero should exhibit extraordinary physical attributes. And so, after Lenin's death in 1924, an Immortalisation Commission was established with the task of taking all necessary steps to preserve his entire body, with special attention paid to the brain. This was kept for two years in formaldehyde before being given to the German neurologist Oskar Vogt, a very prominent figure with some radical ideas on brain structure. He was entrusted with Lenin's brain and allowed to take it back to Berlin.

The brain, which weighed 1.34 kilos, was chopped into four parts, each of which was sliced into 7,500 sections. Vogt found cortical pyramidal cells of unusual size which, it was said, indicated genius. The brain was kept in Berlin through the Second World War and, it is claimed, a special Red Army detachment was sent to return it to Moscow in case the Americans should seize it and, presumably, decipher the secrets of its superiority. However, studies of the brain after the fall of Communism concluded that the enlarged pyramidal cells indicated nothing in particular. Stalin's brain was said to be even less remarkable.

Inspired by Vogt's methods, on 18 April 1955 Thomas Harvey, the pathologist on call at Princeton Hospital, conducted an autopsy on Albert Einstein and took possession of – effectively stole – his brain. This was against the wishes of Einstein himself who had specified cremation and the scattering of his ashes at a secret location to discourage the formation of a cult. Harvey also took Einstein's eyes – 'Clear as crystal,' said his ophthalmologist – which were last heard of in a bank vault in New Jersey. The operation having been exposed as unauthorised, Harvey refused to surrender the brain and as a result lost his job.

At Philadelphia Hospital, he had the brain cut into two hundred blocks – there seems to be an obsession with chopping brains into small pieces throughout this history – most of which he stored in his house. These pieces then travelled with Harvey across America while he made repeated attempts to interest scientists in the stolen organ. One scientist took the bait and concluded that there was something abnormal – an unusually large number of glial cells – about the brain, again supposedly suggestive of genius, but the study was quickly discredited. Other abnormalities were found, but none can conclusively be claimed to account for Einstein's prodigious intellect.

The assumption that there must be some immediately recognisable physical correlate of human superiority within the brain is understandable. Large muscle mass tends to mean great strength so a large brain should mean high intelligence. But superiority is a less easily defined attribute than strength – definitions will vary from culture to culture if not from individual to individual. Lenin was a brutal politician, Einstein a brilliant physicist, Turgenev and Whitman great writers. Which you think is superior is a matter of taste and belief rather than science and is unlikely to be reducible to a single measure like weight or general organisation.

But, then, how do we penetrate the mysteries of the physical brain? The problem for scientists is that, under normal circumstances, they cannot experiment on the living human brain and are forced to rely on the subjective reporting of the brain's owner. The exceptions to this rule are cases like Henry's where definable damage can be linked to observable effects which are in no way dependent on the patient's subjective reporting.

Surgery is one way in which this might happen; another way is accidental damage. Henry's predecessor as an accidental hero of neuroscience was Phineas T. Gage, a railroad worker who, on 13 September 1848, was using an iron rod, three feet seven inches long, an inch and a quarter in diameter and weighing over six pounds, to tamp down some explosive. The charge went off causing the entire rod to pass through Gage's head and land eighty feet away. Gage remained conscious in spite of the loss of large parts of his brain. The first doctor who examined him noted the sight of his pulsing brain in the wound. Gage vomited in his surgery, the effort of which 'pressed out about half a teacupful of the brain, which fell upon the floor'.

Gage's character was affected. Previously known as a well-balanced individual, he became, according to Dr John Martyn Harlow, who attended him after the accident, 'fitful, irreverent' and he 'indulged at times in the grossest profanity'. But, changed or not, he lived for another twelve years. The case showed, first, that survival of such an accident was possible; the brain was unexpectedly robust. In this case Gage was probably saved by the fact that we all have two brains – the right and left hemispheres are, roughly speaking, duplicates and his left hemisphere seems to have been largely unharmed.

Other than that, his role in neuroscience is ambiguous. There are two contradictory views of how the brain works. One is that all its functions are distributed around the entire brain and there are few, if any, areas dedicated to specific functions like language or sight; the other view is that there are many such areas. Gage was used as an example supporting both of these views.

Weighing brains or studying those of Einstein and Lenin were attempts to escape from purely abstract accounts of the human mind and to anchor our thoughts and impulses in the physical substance of the brain because, to the materialist, that is where they *must* be anchored. With Gage, and especially with Henry, that faith began to acquire scientific credibility. With the arrival of fMRI machines, it began to seem routinely verifiable.

Establishing the physicality of the mind by anatomising the brain has certain important implications. If this one or two kilograms of matter contained the human world, then perhaps it could be artificially replicated or contained by other, more convenient means than the human body. This has now become one of the great projects of our time.

In 2001 French archaeologists found cat and human remains buried together in a 9,500-year-old grave in Cyprus. Cats are not native to Cyprus so they must have been brought there by humans, indicating they had already been domesticated. Art depicting cats and even mummified cats, dating back four thousand years, has been found in Egypt. The Egyptians also worshipped feline goddesses. Dogs may be older companions – 12,000-year-old dog/human burials have been found – but cats seem to bear stronger spiritual overtones. It is perhaps their way of never appearing to have fully accepted domestication. They roam at will and, in contrast to the displays of dogs, there is always something reserved about their shows of affection for humans. Cats are mysterious and seem to conceal an inner life.

DARPA is the Defense Advanced Research Projects Agency, a branch of the American Defense Department. It develops technologies for the military. Future wars are very hard to imagine; almost anything may turn out to be militarily useful including better knowledge of the workings of the human brain. DARPA has, therefore, financed an IBM research project aimed at 'reverse engineering' the brain on a computer. In November 2009, IBM announced it had succeeded in modelling a cat-sized brain.

It was the fact that it was cat-sized that gave the story such resonance. Cats are our constant companions, so much so that they barely seem like animals at all but, rather, familiars. Yet they are reticent, they appear to have an inwardness like our own. Here, it seemed, was a cat on a computer; it was a precursor of finding ourselves.

Dhamendra Modha, the IBM project's lead researcher, announced that their 'cortical simulator' running on a super-computer at the Lawrence Livermore Laboratory had generated a digital simulation of a billion neurons connected by ten trillion synapses. This was said to be the equivalent of a cat brain. The previous largest structure generated by the simulator had been equivalent to the brain of a mouse.

Ambitious claims have been made for this DARPA/IBM project; it is said to be brain science's equivalent of NASA's Hubble Telescope or Europe's Large Hadron Collider (LHC), the two most spectacular scientific tools of our time. The Hubble sends back detailed pictures of deep space and has changed our under-standing of the cosmos; the LHC has been built to explore the ultimate nature of the material universe. The IBM project is expected to do something just as spectacular – decipher the workings of the human mind.

IBM is similarly involved in the Blue Brain Project in Lausanne, Switzerland. Using a customised Blue Gene supercomputer, a team led by neurologist Henry Markram is also attempting to reverse engineer the brain.

Markram's team spent a decade preparing for the project by constructing a database of all that was known about the archi-tecture of the neocortex, the outer layer of the brains of mammals. The brain's highest functions – reasoning, thought, language – seem to be controlled in the neocortex so, it is assumed, whatever is most distinctive about human beings is to be found in these six thin layers of grey matter. The key structures are neocortical columns. These are about 2mm long and 0.5mm thick and contain up to 70,000 neurons. Markram describes these as the 'network units' of the brain.

This information alone would not have been enough to launch Blue Brain. Enormous computing power was also needed and it was only when that became available that the simulation programme could begin. Blue Brain was launched in 2005. In 2009 Markram told a TED (Technology, Entertainment and Design) conference that it would take ten years to complete.

The importance of these projects is that they represent an entirely new way of gaining access to the human brain, the most improbable and complex system yet discovered. It is improbable because it is, or seems to be, a system for producing consciousness from matter, to be exact from about 1.3 kilograms of mainly fat and water. It is complex because the processes of consciousness require phenomenal levels of flexibility, speed and memory storage and many other qualities of which, as yet, we know nothing.

Both Markram and DARPA are using a method called reverse engineering. Previously it was thought we could build an intelligent, brain-like machine by constructing ever more powerful computers. By connecting billions of transistors, we would eventually arrive at a machine of true complexity that would learn and think for itself. This was a way of making a brain from the bottom up. Reverse engineering, in contrast, starts from the finished object and works backwards. It is an engineer's rather than a scientist's approach.

These brain simulators would construct a virtual version of the complete brain. They would not be the brain itself any more than your avatar in the computer world of Second Life is you. At our current state of knowledge, this simulation seems much more achievable than previous approaches which were based on the over-simple idea that a brain need not be merely simulated, it could actually be constructed. But, for more than half a century, attempts to build artificially intelligent machines have constantly foundered when confronted with the complexity of the real thing.

The idea of artificial intelligence (AI) was born at a summer school in 1956 at Dartmouth College in New Hampshire. The phrase itself was invented by John McCarthy, a computer and cognitive scientist. Two of the most celebrated scientists at the school were Herbert Simon and Marvin Minsky. Simon predicted that 'machines will be capable, within twenty years, of doing any work a man can do', and Minsky said 'within a generation . . . the problem of creating "artificial intelligence" will substantially be solved'. This was to prove over-optimistic.

The project was always dogged by two conceptual problems. First, even the most banal things humans do turn out to be phenomenally difficult to replicate artificially. Secondly, AI is based on the belief that consciousness will emerge from a sufficiently complicated machine. This has come to seem less and less likely and the idea seems to have been abandoned.

In 2007, Minsky, one of AI's most important prophets, admitted there was something fundamentally wrong with the idea that consciousness would simply emerge from a sufficiently complex machine. 'If mere complexity were enough,' he wrote, 'then almost everything would have consciousness!'

The implication is that we would have to design consciousness into the machine, we could not just wait for it to happen. But we lack any agreed definition of consciousness and it has not sprung into view on our fMRI scans. So, for the moment, conscious software is inconceivable.

AI, therefore, has failed to fulfil expectations, as has the related field of robotics. In spite of all the optimism about robots, the best we have remain almost comically incompetent. Honda, for example, had been working on a line of robots since 1986 when, in 2000, it unveiled Asimo. Asimos cost $1 million apiece and are, indeed, a big advance. An Asimo can detect several moving objects, greet people, react to voice commands and gestures, climb stairs, navigate rooms and distinguish up to ten faces. It is also cute, looking like a child-sized astronaut. But it is a long way from being an intelligent machine.

The search for the 'more' – the spark of consciousness, the glimmer of intelligence – that is missing from our machines is now a very urgent project not just in science but also in philosophy. In fact, it is a project that seems to exist at the intersection of these two usually competing disciplines – philosophy asking the 'what' and 'why' questions, science asking the 'how'. This is a very fraught intersection, with scientists at one end of the spectrum saying it is just a matter of time before we effectively eliminate the need for philosophical head-scratching about the mind, and philosophers at the other end saying we have not even begun to address the central issue which is how matter becomes mind.

Perhaps the most celebrated current idea is that the missing 'more' is the emotions. The Portuguese neuroscientist Antonio Damasio (who now works in America) is the champion of the view that emotions are not some separate add-on to the processing capacity of the brain. Rather, they are embedded in everything we do. Against much of the dominant thinking of the twentieth century, he shows that emotions are involved in even the most seemingly rational high-level decision-making. Damasio credits the seventeenth-century philosopher Baruch Spinoza, also Portuguese, for first glimpsing the contemporary insights of neuroscience on the issue of the mind and the body. Again the intersection.

Damasio's point about emotions is that they are all about the internal state of the brain. Our senses provide contact with the external world, but the emotions are necessary to process the incoming information internally. Emotions are not, as is often said, a barrier to rationality, they are an essential aspect of it. Damasio found that patients with damage to emotional centres of the brain continue to perform well on intelligence tests, but they fail badly in planning, judgement and social awareness.

Some argue that Damasio has in fact produced a persuasive

definition of consciousness. Animals react to sensory stimuli but only humans do so in combination with a sense of our bodies and our inner selves. Consciousness is an awareness of the somatic – bodily – environment and our inner state which allows us to evaluate incoming perceptions, a process Damasio calls somatic marking.

In a way, the role of emotions should always have been obvious. Our emotional state colours everything we do. As the philosopher Ludwig Wittgenstein observed in 1918, long before anybody had thought in practical terms of AI, the world of the happy man is quite different from the world of the sad man. Sadness or happiness are present in every step we take, every word we read.

Less obvious is what we can do with this insight. If our consciousness is so entangled with our emotions, then making a machine that replicates the human mind means it must be an emotional machine. Can such a thing be built or is consciousness dependent on a human body and a human brain?

Our failure to make an intelligent machine or to explain how matter becomes mind, an explanation of *how* the brain manages to be wider than the sky, must mean that materialism remains a faith. We have not yet proved – and we are very far from proving – that the world is entirely material. It is all very well to say the act of writing a great poem is the product of electrical and chemical activity in the brain of the writer, which, in turn, is the product of several billion years of evolution, but there remains an abyss between all that and the apparent immateriality of the thoughts and feelings in the creative mind of the artist.

Sweet-natured Henry Molaison may have been a clue. We know exactly what made him lose his memory. Unfortunately, we still know nothing of what it meant to be Henry. He died, as he had lived, unique.

A SIGNATURE SCIENCE

Henry Molaison lived in a gap, the illuminated moment of an inexplicable present in the midst of past and future darkness. His mind constantly met the world anew. Perhaps his predicament was not so far from that of the poet.

Wallace Stevens spent his entire imaginative life on the strange borderland between the mind and the world. In his finest poems he seems to merge the two into some perfect unity. In 'Not Ideas About the Thing But the Thing Itself', a very late work, he uses the experience called either hypnogogic or hypnopompic – the interval midway between waking and sleeping in which we can become confused about whether something really happened or was imagined. In the poem a bird's cry 'seemed like a sound in the mind' of the hearer. But was it a real cry? In the end it does not matter: 'It was like / A new knowledge of reality'. The world need not be – cannot be – a stable entity outside the mind but in the borderland between the mind and the world lies a way of knowing.

Born in the late nineteenth century, Stevens spent most of his life in Hartford, Connecticut, where he worked for the Hartford Accident and Indemnity Company, eventually rising to the rank of vice-president. *Harmonium*, his first book of poetry, was published in 1923, when he was in his fifties.

He was admired in Hartford but, among his colleagues and neighbours, he was known as somewhat anti-social and prone to abruptness. People were seldom let into his house; visiting poets were shocked to find themselves booked into local hotels. This may have been because of the signs of mental illness in his wife, Elsie, but it was also because of Stevens' commitment to the cultivation of his own thought processes. 'You have to think two or three hours every day,' he wrote to the Cuban poet José Rodríguez Feo. 'You have to think [not only] about what you read, but you have to think about your life and the things around you.'

His sequestration in Hartford and his respectable life echo the quietness of Emily Dickinson in Amherst, Massachusetts. Two of the greatest American poets led, to all outward appearances, ordinary public lives as if to protect the limitless complexity of their imaginations. But their lives were boring in very different ways. Stevens had no shameful disease, he had a stable marriage and he did travel, notably to Key West, Florida. It was in Key West that he displayed his most exotic characteristics, drunkenly rowing with the poet Robert Frost and breaking his hand punching Ernest Hemingway on the jaw, though Hemingway seems to have won the fight.

There was, however, a much more profound connection between Stevens and Dickinson than mere quietness and respectability. They both understood what neuroscience is only just beginning to grasp and what physicists have understood since the publication of Werner Heisenberg's Uncertainty Principle in 1927 – that the place where the mind ends and where the world begins is unknowable.

This makes them both poets of the second machine age. The physical transformation of the natural world in the first machine age, the time of the Industrial Revolution, inspired dismay in poets such as William Blake and William Wordsworth and outright disgust in the great Victorian art critic and philosopher John Ruskin. They saw the destruction of the organic basis of society, the death of the gods and the alienation of the craftsman from his craft. Dickinson and Stevens saw something quite different.

In the second machine age, the challenge to the human world is mental rather than physical. As the gadgets become more intimate and the scanners more powerful, it is our inner worlds that are being transformed. Perhaps they are even being destroyed. The perpetual connection and distraction of our lives now are the opposite of Stevens' solitary thinking time or Dickinson's isolation in her room. Connectivity is replacing creativity on Facebook and Twitter.

It is not that Dickinson and Stevens were protesting against this as Wordsworth protested against the railway invading his beloved Lake District – how could they? For both of them global connectivity lay in the unimaginable future. Rather, they both saw the instability, the mystery, of the mind–world borderland. Specifically, Stevens saw that with the death of the gods our sense of the world had fundamentally changed. The supposedly real had gone to be replaced by the shifting – but, for Stevens, greater – reality of our imaginations. There is comfort in this because, as he says in one of his last poems, it leaves us free to say 'God and the imagination are one'. Thinking quietly, unconnected and with our guests staying at a local hotel, we could find peace.

But can our brains ever again be left in peace? The very inwardness of our minds has become a pressing scientific issue. At one level, this is because the human brain is, as I have said, the one part of the world that has yet to be drawn into the materialist world view. The lack of a clear connection between the mind and the brain is a challenge to science as a completable project and as a subject that must be wrenched from the grasp of the philosophers.

At another level, it is also a primary obligation of science to improve the human condition. Knowledge for its own sake is, perhaps, the highest and purest end of study. But science is too restless and too effective to stay locked in that box for long. If knowledge can improve our material lives, then it must do so. Neuroscience is now seen as the most promising way to effect improvements.

It is the most discussed scientific discipline in the world, superseding physics and biology on the science shelves of bookshops.

Like physics it asks questions about our place in the universe; like biology it examines how we came to be, but, more intimately than both, it probes the human self, how we feel, who we are and how we make a world in our heads.

The possibility of knowing how the brain works has changed the language. 'Neuro' is suddenly a commonplace prefix – neuroeconomics, neuromanagement, neurocapitalism, neuropsychology, neuromarketing. The pictures of the brain in action produced by the fMRI machine have inspired a conviction that we are on the brink of being able to observe, control and exploit the most intimate aspects of our consciousness. What discipline, what market sector could not be affected by such power?

There is a huge consumer aspect to all this. We are living longer and some diseases of old age are being well controlled by medication, such as statins for cholesterol, or ACE inhibitors for high blood pressure. But there is one disease which resists medical intervention.

Brain decay just happens. Your brain is made to last at best about one hundred years – though there are a few exceptional cases – and cognitive decline sets in well before that. Just now, the first signs of decay are setting in for babyboomers, the postwar generation – my generation – now aged between forty-eight and sixty-six. Boomers in the developed world have had a good life, enjoying the fruits of peace at home and often explosive economic growth. Technological innovation has raised confidence in the idea that all problems can be solved.

But the boomers are losing their glasses or forgetting the names of people they have known for years, a phenomenon known as age-associated memory impairment (AAMI). Memory has, as a result, become a babyboomer obsession. A cover story in *Newsweek* in 1998 – just as the boomers were beginning to feel the effects of AAMI – announced that memory loss had become the main health concern of the entire generation. Stress and distraction were said to be making them forget. Daniel Schacter, professor of psychology at Harvard, writes of the boomers 'grumbling in record numbers about their increasing propensity for forgetting'. Schacter, born in 1952 and therefore a boomer himself, helpfully listed 'the seven sins of memory': transience, absentmindedness, blocking, misattribution, suggestibility, persistence and bias. Memory failures are, of course, further evidence of the abyss, the borderland, that separates our minds and the world.

Then there is the emblematic disease of our time, Alzheimer's, which offers boomers, in Schacter's words, 'the prospects of a life dominated by catastrophic forgetting.' It is as if the fate of Henry Molaison awaits us all. Many boomers are seeing their parents coming down with Alzheimer's and they know exactly what it means – a relentless disintegration of the self until, finally, the person becomes a living nobody.

The twentieth century was called the century of the self in a BBC documentary series and Tom Wolfe called the Seventies 'the me decade'. These were periods when the cultivation of the self through the gym, alternative medicine, fashion and various new-age spiritualities became the dominant orthodoxy in the developed world. Now, because of Alzheimer's, a falling birth rate and rapidly ageing populations, the twenty-first threatens to become the century of the non-self.

The fear of, literally, losing one's mind has, at the most basic consumer level, spawned a range of products, all based on the principle of 'use it or lose it'. They also sell on the back of the very boomer obsession with the gym. The brain, it is often said (erroneously), is a muscle and needs exercise.

From Sudoku to *Dr Kawashima's Brain Training* games on the Nintendo DS, the market is now awash with ways in which you can exercise the brain. Their efficacy is dubious. 'There's no empirical evidence that these games produce improvements,' says Nancy Andreasen, neuroscientist and author of *The Creative Brain*; 'saying you spend half an hour a day playing Sudoku and you won't get Alzheimer's or playing any of these brain games and you'll lose less grey matter than somebody who doesn't – well, nobody has ever done that study.'

There are, however, one or two games that seem to have some effect. Susanne Jaeggi, a psychologist at the University of Michigan, has devised a strange and complex game involving sequences of squares on a computer screen. It definitely improves 'fluid intelligence' – that part of your mind that deals directly with raw, new experience.

But the real prize for the boomers would be a technological fix for the brain, just as, for scientists, the big prize would be showing how the brain works, how it produces the mind. The pursuit of this prize has been accompanied by a deluge of strange and marvellous insights that seem to overthrow our sense of ourselves and the world. Indeed, in the lay imagination, neuroscience has taken over from physics – with its black holes, quantum indeterminacy and elastic time – as the generator of strange and frequently unimaginable depictions of reality.

At the heart of the matter lies the revelation that the world in our heads is so radically changeable and so often at odds with what we think is stable outer reality. Thanks to the ever-increasing popular literature on the brain, we now know of bizarre conditions like visual agnosia in which the brain simply cannot make sense of visual inputs – most famously there was the man who thought his wife was a hat. Or there is alexia sine agraphia in which the sufferer loses the ability to read but can, strangely, continue to write. Or there is Anton–Babinski syndrome in which people who have been blinded by brain damage persist in believing they can still see. It is one of a class of conditions known as anosognosia, the denial of illness. This may in turn be an aspect of Wernicke's aphasia in which the sufferer speaks fluently and grammatically but incomprehensibly and then frequently grows angry when he cannot be understood. Or there is prosopagnosia in which patients cannot recognise familiar faces. The equally malign reversal of this condition is Capgras syndrome in which faces are recognised but the patient is convinced that the person is an impostor.

It may be comforting to dismiss such conditions solely as symptoms of abnormal brains – and, indeed, they are usually associated with brain damage – but other phenomena involving supposedly normal brains indicate equally bizarre disconnections from what we take to be reality. The Dunning–Kruger effect, for example, is about the way incompetent people have a tendency to over-estimate their skills to the point where they are completely unable to recognise and correct errors. This, however, seems to be a culturally specific effect. Americans exhibit Dunning–Kruger far more than Europeans, and East Asians seem to suffer from a form of anti-Dunning–Kruger: they actually tend to under-estimate their own competence.

But perhaps the most startling revelation about the brain is that it is, in fact, two brains. The anatomical division of the brain into two broadly identical structures reflects the division of much of the body – we have two legs, two arms, two eyes and two lungs. But, unlike those other features, there is something profoundly odd about the fact that we really are in two minds.

Lateralisation of brain function, as it is called, became apparent through the study of epileptic patients. In the 1940s William Van Wagenen, a neurosurgeon, developed a surgical treatment for the condition which involved severing the corpus callosum, the great rope of nerves that joins the right and left hemispheres of the brain. Roger W. Sperry, a Nobel Prize-winning neuropsychologist, examined these patients and concluded that each was conscious in a quite different way. This was not, at first, an obvious finding. Post-surgery patients seemed surprisingly unaffected, able to continue with their normal lives. But experiments conducted by Sperry and, subsequently, by many others revealed the differences. Sperry described the phenomenon in his Nobel Prize acceptance speech: 'Each disconnected hemisphere behaved as if it were not conscious of the cognitive events in the partner hemisphere ... Each brain half, in other words, appeared to have its own, largely separate cognitive domain with its own private perceptual learning and memory experiences.'

The most spectacular evidence of the division of the brain is, perhaps, alien hand syndrome, which most commonly afflicts people whose corpus callosum has been severed. In this condition, the two halves of the body seem to be at war with each other. One hand may button up a shirt, the other will unbutton it. Though sufferers feel sensation in both hands, one is seen to have, somehow, gone rogue and developed a mind of its own. Some patients regard the hand as not belonging to them at all. They strike it and push it away. In one famous case, reported as long ago as 1908, a woman's hand tried to strangle her and had to be violently pulled away.

Social embarrassment is often involved when the alien hand behaves outrageously in company. In fact, divided brain embarrassment can be manifested in other ways. In one Sperry experiment, a split-brain woman was shown a pornographic image. Her right hemisphere made her giggle but her left hemisphere was unable to explain why.

It is usually the left hand that is alien. In most people the left side is controlled from the right of the brain and vice versa. But the sense of self seems to reside in the left brain so the 'I' which takes offence at the behaviour of the hand is, in fact, the left hemisphere judging the actions of the right hemisphere.

These are the observations of contemporary science but the idea that our apparently single mind is, in fact, double is old and has always been possessed of imaginative potency. The concept of the doppelgänger – literally 'double walker', meaning a ghostly presence accompanying an individual – goes back to ancient Norse and Egyptian mythology. Significantly, the doppelgänger, like the rogue hand, is usually evil and disruptive. In Robert Louis Stevenson's *The Strange Case of Dr Jekyll and Mr Hyde*, the hero contains two diametrically opposed personalities – one, Dr Jekyll, good and one, Mr Hyde, evil. Real people frequently see their own double. Abraham Lincoln, for example, was disturbed by the sight of a second image of himself in a mirror. Poets are often afflicted. John Donne saw his wife's double, Shelley saw himself and Goethe saw Goethe riding towards him.

In a curious echo of the power of the fMRI machine, it has been found that magnetic fields can induce the presence of the double. In 2006 it was reported that magnetic stimulating of a patient's brain to treat epilepsy had resulted in her strong feeling that there was another person in the room. Michael Persinger, a cognitive neuroscientist, has put subjects inside an intense magnetic field and induced the sensation of another person in the room. This may explain phenomena such as alien visitations and out-of-body experiences. Persinger explicitly links the effect to brain lateralisation. 'We have suggested that during these transient inter-calations, you become aware of the right hemispheric homologue of the left hemispheric sense of self, and you experience the presence. The presence is effectively your right hemispheric rep-resentation of the self.'

So is our double sense of ourselves the product of brain damage or a wiring glitch between the two sides of our brain to be exposed by powerful magnetic fields? Or is it much more fundamental than this? Is it the way we construct our world?

At this point the double brain story comes back to that Wordsworth poem, 'The world is too much with us'. In that sonnet, the poet longs to see and hear gods in the world as the ancients did. What, neuroscientifically, he may want is a world of the split brain.

The influential American psychologist Julian Jaynes, who died in 1997, thought that human consciousness is primarily defined by meta-awareness: we can think about thinking, observe our own thought processes. This kind of consciousness, he suggested, did not exist until about three thousand years ago. Humans lived unconsciously, following habit and instinct. They had a 'bicameral' (two-roomed) mind. In this mind the left brain's voice of authority was heard as the voice of a god or leader and was to be obeyed instantly.

Schizophrenics who hear voices are effectively reverting to this earlier state. Jaynes argues that, for example, in the older parts of the Old Testament there is no sign of introspection. This only begins to appear later in the Bible and in works such as Homer's *Odyssey*. The pressures of social collapse and mass migration, Jaynes speculated, may have caused the development of consciousness. In the midst of chaos, the gods could no longer be heard so prayers were said to summon them. The two minds merged – or seemed to merge – into one, possibly with the widespread adoption of writing which encouraged the kind of introspective analytical thought we now regard as normal.

The climax of the modern, merged mind is Shakespeare's *Hamlet*. The literary critic Harold Bloom credited Shakespeare with inventing the human. By this he meant the process of introspection, of knowing the voice in your head is not that of a god but your own. The character of Hamlet is magnificently introspective; he speaks some of the greatest poetry ever written. But his introspection also tortures him. He cannot escape from the crisis in his life – the murder of his father – and he is plunged into long agonies of indecision and driven to thoughts of suicide. He is also phenomenally intelligent and he sees the significance of his own crisis in terms of the nature of humanity in general.

> What a piece of work is a man, how noble in reason, how infinite in faculties, in form and moving how express and admirable, in action how like an angel, in apprehension how like a god! the beauty of the world, the paragon of animals – and yet, to me, what is this quintessence of dust? Man delights not me: no, nor woman neither, though by your smiling you seem to say so.

Man's capacity for introspection makes him like a god, but he remains 'a quintessence of dust'.

The implication of this is clear: we have lost something – the ability to hear the voices – and this loss has been imposed on us by the analytical modes of the modern world. Hamlet's great soliloquies are expressions of agony at the sheer burden of consciousness. Wordsworth wanted to see and hear again the old gods of the world, even if the belief system that sustained them had become obsolete – 'Great God!' I'd rather / be a Pagan suckled in a creed outworn.'

Iain McGilchrist, a literary scholar-turned-psychiatrist, takes this sense of loss much further in his book *The Master and his Emissary: The Divided Brain and the Making of the Western World.* In this he argues that the left hemisphere of the brain has now utterly taken over our lives. The left hemisphere deals with rationality, rule following and exact ordering of information, the right with more subtle matters such as context, meaning and all the nuances of our interactions with the world.

He first saw the results of this in literary studies, a field in which various quasi-scientific ideas – structuralism, deconstruction – had taken over direct engagement with the work. 'I wasn't very taken with all the "isms" of the time,' he told me, 'which seemed to me in their own way to take on the mantle of some sort of false scientific scheme that would make them credible.'

His book *Against Criticism* confronts the isms head-on. 'It was basically about this whole problem about works of art – we abstracted from them, decontextualised them, treated them as non-incarnate objects and then they suddenly turned out to be a banal collection of ideas you could have found somewhere else. Whereas the whole thing about them seemed to me to be their obstinate, incarnate quiddity and they were in this way like human beings.'

Works of art are as complex as human beings and yet, in our left-brain world, we are determined to reduce them to pseudo-scientific categories. The really strange thing about this idea is,

McGilchrist says, the fact that the left brain has almost no direct contact with the outside world, it orders and categorises rather than feels or experiences. It draws information from the right brain rather than from the world. A left-brain view of a work of art tends to be emotionless and removed from direct sensuous experience.

The right–left division of the brain has penetrated popular consciousness. It has, in recent years, become routine to hear people say 'that's so left-brain' meaning something is cold, rational and uncool. These tend to be over-simplifications of an enormously complex area and McGilchrist himself admits that the idea may be a metaphor, a way of understanding real changes in the world rather than an accurate description of the causes. Larry Parsons told me this strict division of functions is a very strong claim that would not be supported by most neuroscientists.

Whatever the status of the left–right division of the brain as science, it is an expression of our unease in the neuro-era in which we now live. There is some instability, some fundamental uncertainty, in our sense of the relationship between the mind and the world. The mind, it seems, can alter its world picture to the point where we can no longer be sure of our connection to the real.

Meanwhile, from the other side of the mind–world equation, since Heisenberg formulated the Uncertainty Principle, the physicists have been telling us that, even if our minds were perfect recorders of reality, reality itself was not there to be recorded. We do not see a world; we invent it out of the chaos of quantum flux which is the 'real' world that our inventions hide from our eyes. This is a quantum version of an argument advanced in the eighteenth century by Bishop George Berkeley. Berkeley's form of idealism claimed that the world is, in effect, created by mind. *Esse est percipi* was Berkeley's motto – to be is to be perceived.

This is not such an extreme idea. In fact, it resonates with the peculiarity of our own experience. Who would we be if we were not perceived? It is not clear we would be anybody. As ever, the artists got there first.

Samuel Beckett was a hard man to meet. You could be rendered speechless by his astounding, hawk-like features, his eyes like Sapo's in his novel *Malone Dies* – 'pale and unwavering as a gull's' – and his limitless sensitivity. Every pain, every thought, every wrong word seemed to register instantly on his features to the point where you could be actually lost for words – exactly the right turn of phrase in Beckett's case since everything he wrote tended towards silence and sadness. 'I had,' he once said of his childhood, 'little talent for happiness.' Yet he was one of the funniest writers who ever lived.

This may explain why the one film he made, called *Film*, starred Buster Keaton. Beckett originally wanted Charlie Chaplin, but somehow he ended up casting the then sixty-eight-year-old Keaton. When the director, Alan Schneider, found Keaton, he was alone and ill and apparently $2 million up in a four-handed poker game with an imaginary Louis B. Mayer and two other equally imaginary Hollywood bosses. Schneider said the game had been going on since 1927; *Film* was shot in 1964.

It is silent and runs for just twenty-four minutes. There are two main characters – O, played by Keaton, and E, which is the camera. O is in flight from E because he cannot bear to be seen or perceived in any way. Other minor characters look directly at E and their faces fill with horror. Once in his own room, O carefully covers a mirror and every possible eye including those of animals – a cat and a dog are ejected and a parrot's cage and a goldfish bowl are covered. He sits in a rocking chair and goes through old photographs, examining them and then tearing them up. He dozes. E gazes at him from different angles, finally frontally. O wakes, stares at E, his face filling with horror. Then we see E – it is O staring back at him. Like Goethe, Shelley and Lincoln, he sees himself as another.

Film is a variation on Berkeley's 'to be is to be perceived'. O wants to escape existence and to be seen is to exist. In the end, however, he must endure existence because he is seen by himself. That, in the second machine age, is precisely the predicament in which we find ourselves.

COUNTDOWN TO THE SINGULARITY

In spite of the multiple failures of the artificial intelligence project, the continuing mystery of a mind as closely studied as that of Henry Molaison, the poor resolution of our brain 'telescopes', and in spite of our own misgivings and the insights of our greatest artists, a defining feature of the world in which we find ourselves is the widespread conviction that the hard problem of consciousness is on the brink of being solved. Or, at least, it is on the brink of being taken off our hands by explosive technological progress.

Ray Kurzweil excites extremes of admiration from the brightest and most powerful. He is, says Bill Gates, 'the best person I know at predicting the future of artificial intelligence' His book *The Singularity Is Near: When Humans Transcend Biology* was an international bestseller which established him as the world's leading prophet of the technological future. The *Wall Street Journal* described Kurzweil as a 'restless genius' and *Forbes* magazine called him 'the rightful heir to Thomas Edison' and 'the ultimate thinking machine'. He has founded eleven companies. His inventions run from a computerised expert system for music composition in 1964, when he was sixteen, to a print-to-speech reading machine for the blind in 2006.

Kurzweil is also Chancellor of the Singularity University (SU), established in 2009 and based in the huge NASA Ames Research Center in Silicon Valley, south of San Francisco. Ames develops the technology for NASA's space missions and it lies next door to the Googleplex, the headquarters of Google. Both NASA and Google are backers of the SU. The goal of the SU is 'to assemble, educate and inspire leaders who strive to understand and facilitate the development of exponentially advancing technologies in order to address humanity's grand challenges'. The SU is not a full undergraduate university; instead it offers summer schools and a graduate programme in subjects ranging from AI and robotics to Future Studies and Forecasting.

Kurzweil is one of the most influential and energetic thinkers in the world today and he is backed by powerful people and institutions. He has a gripping story to tell of the future and he tells it with phenomenal drive and commitment. It is the story of the Singularity.

The idea of the Singularity was first floated, apparently casually, by the mathematician and computer scientist John von Neumann in the 1950s. 'One conversation centered,' ran a report of von Neumann's words, 'on the ever accelerating progress of technology and changes in the mode of human life, which gives the appearance of approaching some essential singularity in the history of the race beyond which human affairs, as we know them, could not continue.'

Following von Neumann, in 1993, Vernor Vinge, a mathematician, computer scientist and science fiction author, published an essay entitled 'The Coming Technological Singularity'. 'Within thirty years,' Vinge said, 'we will have the technological means to create superhuman intelligence. Shortly after, the human era will be ended.' He calls this moment the technological singularity, a term derived from physics. Before Vinge wrote this essay, a singularity was most commonly used to denote what happens at the centre of a black hole or at the beginning of the universe. This may be called a gravitational singularity in which matter collapses on itself, becoming infinitely dense. What happens inside a singularity is unknowable because, at the point, the laws of physics no longer apply. Similarly, after Vinge's singularity the human era will be over and our present ways of life will no longer apply.

'The acceleration of technological progress,' he wrote, 'has been the central feature of this [the twentieth] century. I argue in this paper that we are on the edge of change comparable to the rise of human life on Earth. The precise cause of this change is the imminent creation by technology of entities with greater than human intelligence.'

This superhuman intelligence will be as inaccessible to our ordinary minds as the centre of a black hole. We will be in the presence of a machine that can, intellectually, leave us far behind. Vinge acknowledges that some will see this as an unwelcome prospect: 'From one angle, the vision fits many of our happiest dreams: a place unending, where we can truly know one another and understand the deepest mysteries. From another angle, it's a lot like the worst case scenario.'

The logic of the Singularity is persuasive. If technology continues to advance exponentially, then it seems possible that we would reach a take-off point, a moment when the technology becomes autonomous, escaping from the limitations of the human brain. As Kurzweil puts it: 'The Singularity is an era in which our intelligence will become increasingly nonbiological and trillions of times more powerful than it is today – the dawning of a new civilization that will enable us to transcend our biological limitations and amplify our creativity.'

He expects the Singularity to occur around 2045. At this point the era of biological intelligence will end to be succeeded by the era of machine intelligence. This may sound radical, perhaps deranged, but Kurzweil is no fringe figure, no hellfire preacher. His speaking manner is that of a successful lawyer addressing a legal conference and his style – the clothes, the hair, the glasses – are the same.

The style signals what may seem an unlikely truth about Kurzweil – he is a deeply conventional thinker. His belief in exponential technological progress is simply a well-organised version of the most pervasive contemporary orthodoxy. Most of us assume, either consciously or unconsciously, that our gadgets will relentlessly improve and that science will continue to advance. All that Kurzweil adds to this assumption is the conviction that these advances will be exponential, leading to a point where advances that would have once taken a century happen in an hour or less.

In October 2010 Kurzweil travelled to Brussels, Zurich, Warsaw and Vienna, lecturing academics, businessmen and politicians. In a speech, entitled 'The Acceleration of Technology in the 21st Century: The Impact on Entrepreneurs', he asked the audience four questions at the beginning and the same four questions at the end to find out how many minds had been changed. Here are two of the questions and the results:

When will computers pass the Turing Test (that is, when will computer intelligence be indistinguishable from human intelligence)?

	BEFORE	AFTER
Within 10 years	31%	21%
Within 25 years	40%	60%
Within 50 years	7%	1%
Within 100 years	1%,	3%
Never	21%	15%

When will we have desktop 'printers' that can 'print' any possible three-dimensional object (such as a blouse, a solar panel, a module to build houses)?

	BEFORE	AFTER
Within 10 years	53%	57%
Within 25 years	18%	30%
Within 50 years	11%	3%

Within 100 years	11%	4%
Never	7%,	6%

The results show that Kurzweil is persuasive: people become more bullish about technology after hearing him speak. But, more importantly, they show that very rapid technological progress is taken for granted. Before he spoke, 71 per cent of the audience thought computers would become intelligent within twenty-five years, and the same number expected all-purpose 3D printers within twenty-five years. The figures increased, but they were high already.

The audience may not have explicitly said they believed in the Singularity, but, implicitly, they did. In spite of its sci-fi, apocalyptic associations, it has become no more than the predictable climax to what is seen as our unstoppable upward trajectory.

This is, in fact, precisely how it is seen by some philosophers, for it is not just businessmen and technocrats who are persuaded by this vision. The belief that we shall – and should – transcend our biological condition is known in philosophical circles as transhumanism. Nick Bostrom, a Swedish philosopher at the University of Oxford, is the co-founder of the World Trans-humanist Association. He defends transhumanism not as an end of humanity, as some of its technocrat followers seem to believe, but as its logical extension.

'For as long as I can remember,' he says, 'it seemed clear to me that the most decisive factor in determining the future of human kind would be new technological developments that could radically change the human condition not so much by altering the world around us, but through the use of technology to change human biology, human capacities.

'If I wanted to make some positive difference in the world, this would be the area with the greatest leverage ... It's an extension of humanism. It has its roots in secular humanism and it shares a lot of its principles with humanism – the idea that human beings are relying on their own powers rather than in deference to supernatural intervention.'

It is a seductive and optimistic project and, like many visions of the future, transhumanism and the Singularity, once seen, cannot be unseen. The massive presence of this technological apotheosis in the near future looms over the present and rewrites the past. Technology becomes the medium through which we see all human developments.

On his website, for example, Kurzweil reviews James Cameron's film *Avatar*. He trashes it, largely on the basis of the technology. The film is set in 2154, but he points out that the weapons used by the humans invading the alien planet of Pandora could be from the First or even Second World War. Worst of all, 'The movie was fundamentally anti-technology . . . I got the sense that Cameron was loath to show modern technology doing anything useful.'

The movie is, conceptually, anti-technology as a force for subduing nature and destroying societies in touch with nature, though, paradoxically, it is also the most technologically advanced film ever made and Cameron is a man obsessed with gadgets and high-tech innovation. But Kurzweil's broad complaint is that, like many sci-fi movies, *Avatar* shows the future as the same as the present with just a few changes. For him, the future's primary defining characteristic is that it is *nothing like* the present.

'Most sci-fi films,' he writes, 'depict a few truly clever technologies representing a probable human future, while leaving the rest too ordinary and undeveloped to be believable. The entire world of human technology will evolve in step, affecting all aspects of the way we work, live, play, heal, create, learn or defend. Advanced technology will be embedded everywhere, in even our most mundane objects, interconnected and always-on. In a future world capable of strong AI and inter-stellar travel, the landscape of technology merged with our daily activities will actually be far more advanced, and far more interesting than in the film depictions we see today.'

The Singularity is everywhere and everything. There is something religious about this vision and certainly some moral force in Kurzweil's celebration of the vast, possibly infinite, landscape of gadgetry that awaits us in the near future. Cameron's goals were aesthetic, but, to Kurzweil, his failure is moral.

Kurzweil has his critics, of course. German director Jens Schanze in his documentary movie *Plug & Pray* contrasts the techno-optimism of Kurzweil with the born-again pessimism of Joseph Weizenbaum. Weizenbaum, who died in 2008, was one of the most distinguished computer scientists of his generation and the creator in 1966 of a programme called ELIZA which was good enough to convince people they were holding a real conversation with a computer. But he became a dissident, criticising the utopian fantasies of his colleagues. 'It is disastrous,' he says in the film, 'that most of my colleagues believe that we can create an artificial human being. This immense nonsense is related to delusions of grandeur. Maybe if I had known back then what I know now then I'd have said I don't like being in this branch [of science].'

P. Z. Myers, meanwhile, a combative blogger and a biologist at the University of Minnesota, is incensed by Kurzweil. 'There he goes again,' Myers writes, 'making up nonsense and making ridiculous claims that have no relationship to reality. Ray Kurzweil must be able to spin out a good line of bafflegab, because he seems to have the tech media convinced that he's a genius, when he's actually just another Deepak Chopra for the computer science cognoscenti.'

The huge increases in machine intelligence that will occur at the Singularity are, for Kurzweil, the heart of the matter because they will make anything possible, but Myers' point is that the simple extrapolation of computer-based models of the brain to a thinking machine is wrong-headed. The brain may not be like a computer or, even if it is, it may not be like any of our current computers. Neuroscientists are now in pursuit of the 'neural code' that may be the brain's equivalent of the machine code that provides the basis of computer operations. Little is yet known about the neural code, but what is known points to it being complex beyond anything we can yet imagine.

Kurzweil relies on the exponential growth of our knowledge to overcome all such difficulties. Of course, we can't imagine the solutions, but who in 1960 could have imagined fMRI machines or superfast computers you could rest on your lap?

The problem, at least up to now, has seemed to be one of mere scale. Within the human brain there are said to be between 100 and 500 trillion synapses. A synapse is the point at which two neurons, nerve cells, interact. The interaction – call it a firing – may be chemical or electrical. This is the most basic operation within the brain. One cubic centimetre of human brain is capable of 40 trillion synaptic firings per second. The whole brain can handle about 10,000 trillion firings per second.

The equivalent of a synaptic firing inside a computer are FLOPS. FLOPS are 'floating points operation per second'; they are the basic mathematical unit of computation. A teraflop is a trillion flops so the power of the human brain in computer language is 10,000 teraflops, or 10 petaflops. The IBM Roadrunner supercomputer is currently capable of 1.5 petaflops. It was built for the US Department of Energy to simulate the ageing of nuclear materials and check on the safety of the American nuclear arsenal. The Cray Jaguar – also a DoE machine – has reached 1.75 petaflops. The Chinese Tianhe-1A runs at 2.5 petaflops; it is used in oil exploration and aircraft simulation. IBM's Blue Waters project is aimed at producing a computer capable of at least 1 petaflop with a peak speed of 10 petaflops. New technology now being developed by IBM suggests that a supercomputer capable of running at one exaflop – 1,000 petaflops – should be operational by 2018. In terms of sheer speed, that machine would be a hundred times faster than the human brain. The machines would appear to be catching up and overtaking.

This process was entirely predictable. In 1965 Gordon Moore, the co-founder of the chip maker Intel, published an article titled 'Cramming More Components onto Integrated Circuits'. He forecast that, for at least the next ten years, the number of transistors that could be packed into an integrated circuit would double every year. Moore later modified this to a doubling every two years. This became known as Moore's Law and it has proved surprisingly accurate for much longer than Moore expected. Computing power has increased, and continues to increase, exponentially. This is why we will achieve some sort of 'human equivalence' so quickly.

Yet it is inconceivable within this time frame that a computer will be able to think for itself or to match in any way the workings of the brain. This is because Moore's Law is only about hardware, not software. In fact, software seems to obey a kind of reverse Moore's Law. Software just gets bigger and more bloated with features to use up all that extra hardware power. Microsoft Word, for example, is no closer to intelligence than it was ten years ago; it just does more things and soaks up more memory and processor time.

Hardware needs to be told what to do, so, without software, hardware cannot work. This makes the phrase 'human equivalence' ambiguous. If it simply means the hardware will match the human brain, then it means very little. We will pass this milestone and nothing will change. If it means we have achieved a software–hardware combination that will work like a human brain, then it will mean a great deal.

Writing the software requires far more knowledge about how the mind works than we currently have. Software is written in machine code which consists of an endless series of zeros and ones. Cracking the neural code of the human brain is a very distant prospect.

This may not matter. We could write software in machine code that created some new, non-human basis of intelligence that did not require access to the neural code. Or supercomputers in ten or twenty years' time could make themselves conscious. This is a familiar sci-fi story. In, for example, the *Terminator* films the supercomputer Skynet attains self-awareness at 2.14 a.m., Eastern Time, 29 August 1997. The exactness of the timing, delivered in the Austrian-accented monotone of Arnold Schwarzenegger's robot, is significant. It suggests there is a clear tipping point where the complication of the machine falls spontaneously into the complexity of consciousness. This represents the strong AI view that we do not have to actually build machine consciousness, it will simply happen when our machines are sufficiently complicated. Few people still cling to this view.

So perhaps this is all a sci-fi story and we will never make a thinking machine. This would not be all that surprising. If there are no other intelligent beings in the universe – and there are, so far, no signs of any – then that means that intelligence has only happened once in the 13.7 billion years the universe has been in existence. It took all this time and space to make one small creature that thinks. It seems unlikely that this small creature, however clever, would be able to repeat the trick in a few decades.

But, in fact, many people – certainly most technocrats and computer scientists and, as Kurzweil has shown, most businessmen and politicians – believe that is exactly what we will do. The Singularity is the logical outcome of the accelerating technological innovation and of the sudden presence of computers – machines that at least appear to think – in every aspect of our lives. It is also inspired by Moore's Law, the idea that there is something inevitable and law-like about our progression to ever more sophisticated machines and the assumption that this must *logically* happen.

But there is an older, better story that lies at the root of the contemporary acceptance of the idea of the imminent arrival of the human or superhuman computer. Alan Turing was the British mathematician who deciphered the code of the Enigma machine that was captured from the Germans in the Second World War. Some estimate that, by doing so, he shortened the war in Europe by two years. Turing's homosexuality was regarded as a mental illness at that time and he was given hormones to 'cure' him. He killed himself with a cyanide-laced apple in 1954, just before his forty-second birthday.

In 2009 Gordon Brown, the British Prime Minister, apologised for his treatment: 'on behalf of the British government, and all those who live freely thanks to Alan's work I am very proud to say: we're sorry, you deserved so much better'. He certainly did, but at least he lives on in every computer in the world. Turing invented an imaginary device called the Turing Machine which was to become the basis of all modern computers. The machine read symbols one by one from a strip of tape and reacted according to a set of rules. It had an internal 'state' which determined which rules to apply. There was a starting state – a programme – and ensuing states were determined by what symbols had previously been read and in what order. A Turing Machine could, it seemed, do anything including imitate the behaviour of any other Turing Machine, in which case it became a Universal Machine, a device that could calculate anything that was calculable. The implications of this device are bewildering and are still discussed.

More accessible to non-mathematicians was yet another Turing invention – the Turing Test. This was designed to find out if a machine is intelligent. Again, like the machine, this is imaginary, a thought experiment, but, again, it has enormous and still debated implications.

In one room is Brian, in another is Clare. Clare is a computer, Brian a human being. Alice, another human, is in a third room. Alice has been told to talk to Brian and Clare for five minutes. She cannot see them and she only sees their words on a screen or, in the original conception, on a teletype machine. After five minutes she has to decide which is the computer; if she cannot tell the difference, then we would have to say the computer was intelligent.

This is more than a game; it is a drama in which the motives of Brian the human and Clare the computer are different and in conflict. Brian wants to help you guess correctly and decide that Clare is the computer. His best strategy, thought Turing, was to answer truthfully. Clare wants to mislead you by convincing you that she is human, not a computer. She will want to provide a perfect imitation, a falsehood. Brian is an embodied truth-teller; Clare is a disembodied liar.

But note that – and this is crucial for the development of artificial intelligence research – neither is present in the room. Brian's embodiment has been erased as a factor in the equation. You are to judge him solely by his responses on a screen. This obviously gives Clare a chance of winning since her machine body is concealed. It gives her a further advantage in that Brian is denied the use of all the visual cues which humans use to communicate. The test separates intelligence from embodiment and, in doing so, established the imaginative climate in which a thinking computer became not merely plausible but likely.

Writing in 1950, Turing estimated that, by the end of the century, an average Alice would often be outwitted by the machine. By then, he wrote, 'general educated opinion will have altered so much that one will be able to speak of machines thinking without expecting to be contradicted'.

He was, it turned out, wrong. We still cannot make intelligent machines that would regularly survive questioning. But his test was brilliantly clear and it remains the gold standard – perhaps the *only* standard – by which we judge machine intelligence, though people still argue about what exactly it means.

Passing the Turing Test was always going to be very hard. The physicist Frank Tipler, for example, worked out the energy required for someone to 'hand simulate' passing a Turing Test.

This means physically looking things up in books in response to questions from the tester. These things would not just be facts, they would be grammar, tone, all the things involved in human comprehension. Tipler's answer was the equivalent output of three hundred million nuclear power stations.

Yet it is generally assumed that machines will pass the test when they become intelligent. There may be a serious flaw in this logic. Jaron Lanier, a computer scientist – one of the inventors of virtual reality – and a musician has pointed out that the Turing Test can not only be passed if machines get smarter, it can also be passed if humans get less smart. This may seem whimsical but it makes an important point about the test: we are being tested as well as the machine. This is a very important shift of perspective. Lanier was pointing to the possibility that we might start to accept the limitations of the machine.

There are many other arguments about the test, but, in essence, the real shape of the idea is the familiar one of the duck. If it walks like a duck and quacks like a duck, it is a duck. What passing the Turing Test means is, if it talks as if it is intelligent, it is intelligent.

But, surely, it might still be just a machine running very clever programmes and how could that be said to be intelligent? Isn't there something missing? This is the point made by a very incisive thought experiment devised by the philosopher John Searle and published in 1980.

Searle was a sceptic of the claims made by strong AI enthusiasts about the possibility of an intelligent computer. He made the point by asking us to imagine a computer that seems to understand Chinese. Chinese characters are fed into the machine and it responds with Chinese characters appropriate enough to convince a Chinese speaker that he is, in fact, talking to a human. This machine would seem to have passed the Turing Test.

But then, he says, imagine a non-Chinese-speaking man in a room with a book that explains the programme in English. He is fed Chinese characters, looks up appropriate responses – in the form of characters he cannot himself understand – and then passes them out. This human in a box would seem also to pass the test, even though he does not understand a word of Chinese.

The two situations are the same – in the first a machine is following a programme, in the second an uncomprehending person is following the same programme. There is no reason, therefore, on the available evidence, to credit either with an understanding mind.

On the other hand, what evidence do we have for the existence of minds other than our own? This is the real challenge posed by the Turing Test. Note that Turing was not saying a machine that passed his test definitely had a mind, just that we would have no choice but to say it has. There would be no other criterion. Searle implies that there is. But what could this criterion be that could not be met by a sufficiently complex computer programme? The answer is inwardness, our sense of ourselves, our ability not to know but to know that we know. This lies beyond the scope of Turing's Test.

It is no wonder that Kurzweil used the Turing Test in his questions at that conference in Zurich. It is a clear and dramatic demonstration of what is involved. But it leaves open the question of the desirability or otherwise of the advent of machine intelligence. Kurzweil himself has been ambivalent about this. 'Although neither utopian nor dystopian,' he has said, 'this epoch will transform the concepts we rely on to give meaning to our lives, from our business models to the cycle of human life, including death itself.'

In his lectures and talks, he is also plainly aware that most people's visceral response to the machine takeover is disgust and fear. Why should we welcome an event that, in Vinge's terms, represents the extinction of our species?

Science fiction has embedded the idea of the anti-human machine deep in the popular consciousness. In the *Matrix* and

Terminator movies, the reign of the machines results in either a hellish system of control and exploitation or a war of extermination conducted against humans. The point in both cases is that there is no reason why the machines should care about people. In *The Matrix* the machine avatar, Agent Smith, goes even further. He feels positive disgust at the spectacle of the human world.

Evidently, if humanity is to survive, we have to seek protection from the machine future. In 1942 the sci-fi writer Isaac Asimov famously laid out the three laws of robotics that should be built into future machine intelligences:

A robot may not injure a human being or, through inaction, allow a human being to come to harm.

A robot must obey any orders given to it by human beings, except where such orders would conflict with the First Law.

A robot must protect its own existence as long as such protection does not conflict with the First or Second Law.

But these would be entirely ineffective in containing the autonomous, super-intelligent machine that would emerge at the Singularity. It could simply decide the laws were irrelevant; it is highly likely to decide they are stupid.

Eliezer Yudkowsky was inspired by Vernor Vinge's paper on the Singularity. In 2000 he became one of the three co-founders of the Singularity Institute for Artificial Intelligence in Palo Alto, California, the heart of Silicon Valley. It was 'a nonprofit research think tank and public interest institute for the study and advancement of beneficial artificial intelligence and ethical cognitive enhancement'. The following year Yudkowsky produced his own paper entitled 'Creating Friendly AI 1.0: The Analysis and Design of Benevolent Goal Architectures'. The paper outlined a way of ensuring that the artificial super-intelligence would be friendly. It was a way of preventing the sci-fi scenario in which the machines took over and abolished or enslaved us.

He is in a hurry because he thinks, along with Kurzweil, that the Singularity may be near and he wants to be part of what he has no doubt will be 'the most important event to come along in the last few millions years or, hey, why not ever?'

Yudkowsky uses a brief thought experiment to explain how this might be done.

'Are you likely,' he asked me, 'to take a pill that makes you want to kill babies?'

The machine can repeatedly reprogramme itself; the trick, therefore, is to set moral limits to these transformations so that killing babies would be as inconceivable to it as it would be to me. The problem with this is that some human beings do kill babies and, given the history of human conflict, perhaps a majority are capable of doing so in suitably extreme circumstances. Yudkowsky was undeterred by this point.

If, as its enthusiasts believe, the Singularity is inevitable, however, then we have already laid the foundations of the machine future. It is unknowable whether, as Nick Bostrom claims, this will be an extension of the human – and the humane – or whether it will be the beginning of an entirely new, post-human world with entirely alien values. The latter was certainly Vinge's belief; his singularity signalled the end of the human. Kurzweil and others are unclear about this, but it is hard to see, if they are right, how what we now take to be humanity could survive the transition.

The last machine we will ever build will be able, in theory, to boot itself into successively higher levels of intelligence and solve all the problems of our present condition – answering the outstanding questions we have about the nature of matter and of consciousness, curing our diseases, rendering us medically immortal, painting our masterpieces and writing our poems. The brain as we now know it will become a relic. The human era, as

Vinge said, will have ended. On the day the machine takes over, the brain of amiable Henry Gustav Molaison and the results of my scan will become obsolete, mere memories.

HITTING ZERO

We must, it seems, be mere supplicants before the alien machine being that emerges at the moment of the Singularity, begging it to be nice to us and praying that it will allow something of the human to survive. But, until that day, we must endure the shortcomings of the lesser machines with which we already live. These machines – starting with the telephone and ending with the various net-worked devices that have insinuated their ways into our daily lives – exert a twofold pressure. On the one hand, they seduce us, we want them to contain, include and involve us; on the other hand, they demand that we become more 'machine readable'. We pay for inclusion and involvement by becoming more like machines. This has, so far, proved to be an awkward, fractious deal.

Almost daily, we encounter call trees. When you call an organ-isation, a voice answers and offers you a series of options, you choose one and then it offers you further options. Operationally banal yet technologically sophisticated, call trees say a great deal about human–machine interactions. They are, in their very ordin-ary way, a vision of the machine future. This is a future that both humans and machines are building and the process demonstrates the difficulties involved in constructing that future without recon-structing humans.

Call trees are a way of simplifying human callers so that they can be understood by the machine. The options queues do not offer routes to all the answers you might want, but only to those the machine can provide. Nor do they, except in some recent, highly advanced systems, take into account your tone of voice, mood or personality. As a result, they are disliked. One poll suggested Americans thought call trees were the second most irritating thing in their lives after hidden charges.

There is a website called gethuman.com. Gethuman is a movement that 'has been created from the voices of millions of consumers who want to be treated with dignity when they contact a company for customer support'. It began as a single page called 'The IVR cheatsheet'. IVR stands for interactive voice response, otherwise known as automated attendant, automatic call distributor or, more colloquially, call tree.

Gethuman was set up to defeat call trees. It provides tricks to make the computer connect you to a human. So, for example, at the time of writing the site advises calling the Apple number 800–275–2273 and ignoring all the messages. Instead, the caller should just keep hitting zero and he will get through to a human being. Repeatedly hitting zero works in the majority of cases. FedEx, however, requires a little more determination – dial 800–463–3339, say 'agent' immediately and then say 'no'.

Call trees are often very badly designed. Most machines put the caller in a position of deep uncertainty. The options offered may not seem quite right and the caller fears finding himself stuck at the end of an options queue with no way of getting back without starting the whole process again. Uncertainty also springs from the lack of information about how long the process is going to take. You cannot expect to engage with a tree just before an urgent appointment. Trees thus create little wells of dead time.

Then there are the attempts by the software designers to soothe your nerves about the whole process with assurances that this is all being done 'so that we may better direct your call'. Or, most commonly, there is the empty flattery of 'Your call is important to us'.

In contact with a call tree, people sense a kind of twofold manipulation. The most obvious manipulation is the way the options menus try to identify you in a way that is readable to a machine. The second manipulation lies in the transfer of inefficiency. Call trees are machines for transferring working time to the customer. You do all the business of 'directing' your call, not the machine, and you do all the waiting. This, in the worst cases, also means that if you press the wrong option button and do, finally, get through to a human but the wrong one, then it is your fault not theirs. This is one of the more troubling aspects of computerisation in general. Systems become autonomous beings in the corporate imagination and relieve employees of responsibility when things go wrong.

Call trees, banal and routine though they have become, are a defining technology of our time. They are conceptually linked to the creation of modern computing in the mind of Alan Turing. After all, a call tree is a clumsy attempt to get a machine to pass the Turing Test by convincing the caller it is more than just a machine. They are also early expressions of the two-way pull – the humans that want machines and the machines that want humans to be more like them – that, in more sophisticated devices, is becoming the primary drama of our civilisation.

Greg Riker, whom I met at Microsoft in 1994 working on advanced consumer products, was a hippie with long hair and a slightly quizzical, judgemental manner. He walked very quickly in a way that implied not that this was his natural gait but that it was a good thing to do. He strode about the campus with a bum bag containing a Psion organiser – then the state of the art – and a tiny digital voice recorder. On these devices he logged every fleeting insight, thought, idea, appointment or name. He insisted that, at all times, his hands were free because, he pointed out, the sheer amount of brain processing power taken up by just carrying something is enormous.

He was living in advance the life of what he called the wallet PC, a machine that seemed at the time to be in the distant future. It was a small box that contained the contents of your wallet and much more. It would, for example, render credit cards obsolete and become, as it developed further, the only object you needed to take with you. Best of all, Riker argued, it would augment the brain.

The human brain cannot easily carry and make quickly accessible large amounts of details – telephone numbers, addresses and so on. Computers, however, can retain, organise and file limitless quantities of such information. With the wallet PC, he forecast, people would input all the time and their every fleeting impression would be filed to be retrieved at will. If a boy started using these things at the age of, say, eight, he would be able, throughout his life, to call up every thought he had ever had on any subject. Incubated by such machines, Riker said 'a new species' would emerge.

This has happened, though not in quite the systematic way envisaged by Riker. In the ensuing years personal digital assistants (PDAs) grew ever more sophisticated. Then they merged with mobile phones to become smartphones. These in turn connected to the computing 'cloud', a central location from which information could be acquired and through which it could be synchronised. So now, if we enter a date or address on our phone, our laptop or our desktop, it automatically syncs with the other devices.

Such developments have made Riker's ideas about the wallet PC seem quaint. *Of course* we can do all that; the only outstanding issue is whether we do it with an Apple smartphone or one that uses Google's Android software. Yet, before smartphones existed, Riker's dream of uploading a life to a machine seemed wildly improbable, exotic and disturbing.

In fact, uploading a life was only one of three linked ideas that lay behind Riker's enthusiasm. The first was that the technology would effectively create a new species, something superhuman, post-human or just better. The second was the machine augmentation of the brain and the third was the upload itself. This would all be achieved through technology, gadgets.

These ideas had then – and still have – a familiar science-fiction ring. The augmented human, partially united with the machine, evokes the chilling concept of the cyborg, a part biological, part artificial being. This union of the organic and the engineered is commonly seen as frightening. The Borg in *Star Trek* is a species made cold and savage by the organic-machine union and the evil Darth Vadar in *Star Wars* is Anakin Skywalker kept alive by machinery. But, in truth, the cyborg is now the ideal to which all our most advanced technology is tending.

The term was first used in 1960 in a paper entitled 'Cyborgs and Space' by Manfred E. Clynes and Nathan S. Kline. The paper exactly anticipates not only the idealism of Riker but also the contemporary orthodoxy of much of Silicon Valley. This is that current technology is not only world transforming but also human transforming. People no longer talk about gadgets as 'labour saving'; they talk about them as life changing.

Clynes and Kline's paper anticipated this idea through their thoughts about how we would survive space travel. 'Space travel,' they wrote, 'challenges mankind not only technologically but also spiritually, in that it invites man to take an active part in his own biological evolution.'

They were addressing a looming practical problem – the paper appeared the year before Yuri Gagarin became the first human in space. It was at the time assumed that this was the first step on the road to the stars. But humans, in anything like their present form, are plainly not up to the task. They need to take a heavy support system of food, water and air with them, and they are constantly at risk from cosmic radiation and the muscle-wasting effects of living for long periods in zero gravity. These problems would make astronauts dependent on a mass of high maintenance and costly to launch machinery.

'If man in space, in addition to flying his vehicle, must continuously be checking on things and making adjustments merely in order to keep himself alive, he becomes a slave to the machine. The purpose of the Cyborg, as well as his own homeostatic systems, is to provide an organizational system in which such robot-like problems are taken care of automatically and unconsciously, leaving man free to explore, to create, to think, and to feel.'

Clynes and Kline did not seem to notice the irony. In space, they were saying, humans must merge with machines in order to be free of the tyranny of machines. The form of the idea captures the shift from the first machine age to the second, from machines outside us, to machines inside.

The one machine the authors propose in the paper is an 'osmotic pressure pump capsule' which would be inserted into the astronaut and deliver metered doses of drugs. This pump would be sensitive to the condition of the body's various systems. Any imbalances or abnormalities would be automatically corrected by a dose of the appropriate drug. Or the drugs would improve human performance. Wakefulness could be ensured, radiation effects prevented and hibernation induced to reduce food consumption.

That so much could be achieved by one pump now seems a quaint, almost Victorian, notion, a sepia snapshot of the old days of technological innocence when all problems were thought to be soluble. The difficulties of space travel, we now know, are far more intractable than anything that could be imagined in 1960. Cosmic rays and solar flares, cardiac atrophy and arrhythmia and all the other effects of zero gravity are among the still unsolved problems.

Yet, on earth, we now take for granted a high level of integration between man and machine, particularly in medicine. Pacemakers have been routinely attached to people's hearts since the 1970s. Artificial limbs are increasingly machine-like in their use of sensors to adjust to the person's movements. Computers have begun to be used to react to the thoughts of the paralysed. Even kidney dialysis represents a form of cyborgism: a life-sustaining function of the body is taken over by a machine. More ambitiously, the colour-blind artist Neil Harbisson has co-invented the eyeborg, a camera that, in real time, converts colours into sound waves so that the artist can hear colours. And, in his Project Cyborg, Kevin Warwick, professor of cybernetics at Reading University, had an implant in his arm connected to his nervous system. With this, he could connect to the internet and control a robotic arm in Reading from Columbia University in New York.

So the cyborg is the mirror image of our ambitions in robotics and artificial intelligence. We want to make machines intelligent but we also want to, in part at least, turn ourselves into machines. To make this easier, machines like the call tree meet us halfway, politely persuading us to be more like them. This is a two-way street at both ends of which stands the cyborg.

To dismiss this as science fiction is to miss the point because almost all of us are already heavily cyborged. Consumer gadgets have created the most intimate bonds between humans and machines. Nowhere is this more apparent than in the rise of Apple since Steve Jobs returned to the company in the 1996. Through a combination of design and product innovation – iMac, iPod, iPhone, iPad – he rescued Apple from near bankruptcy and made it the most valuable company in America. And he did so by making his machines so desirable, so important and so intimate that, in effect, we had no choice but to adapt ourselves to their demands.

Jobs operates by turning his own fetishistic obsession with gadgets into a global marketing tool. From the beginning his primary drive was not technological but aesthetic. He was once asked what was wrong with his main rival, Microsoft. His answer was that they had no *taste*. Apple's designer, Jonathan Ive, realises Jobs's aspirations in seductive, minimalist forms that owe nothing to the drab cosmetics previously associated with electronic gadgetry and everything to the machine aesthetics of the Bauhaus and Dieter Rams, the German designer for Braun. These are spiritualised, hermetic machines. You are not meant to tinker with them, but, rather, to accept them into your life.

Jobs is capable of holding up products in development for years – as, apparently, he did with the iPad – solely on the basis that they don't satisfy his inner aesthetic ideal. He even insists that the insides of his machines, invisible to the customer, look beautiful. This makes him, in the words of Apple watcher and *Newsweek* writer Dan Lyons, 'the ultimate end-user'. 'Jobs is not an engineer, he can't write code,' he says, 'he can't really design anything and he doesn't know anything about circuits. But he is the ultimate end-user and that makes him the guy who is on our side.'

There is a smooth transference of Jobs's gadget fetish to his customers. The sight of people queuing outside Apple shops to acquire new products has become familiar. Then, caught on TV news, they are shown emerging triumphantly, waving their box. There is also the phenomenon of 'unboxing' in which the gadget fetish becomes explicit porn. Videos are posted on YouTube of new Apple products being taken out of their boxes with each step of the process being lovingly savoured. This may be for people who do not yet have one or it may simply recapture the first hit of new ownership for those who do.

Apple is not, of course, alone in all this. It just does it better than anybody else. What all the companies are doing is engaging us with machines at ever deeper, more personal levels. Above all, thanks to cellphone networks and the internet, they engage us all the time. We are, as John Ashbery said, being renewed by everything. Unlike call trees, these are the machines we love to love and we do not resist them.

The purity, the ethereality, of the design of Apple's machines is an expression of the immateriality of what they contain and produce. They are machines born of abstraction, they manufacture and absorb a non-material entity we have come to know as 'information'. In 1994 when Bill Gates spoke of the Information Superhighway, many people were still puzzled by the idea. I remember people asking me what he meant by 'information'. This was because, in common usage, the term was still narrowly defined – information meant flight schedules, telephone directories or TV listings and there was no reason to think such things needed a superhighway.

But the discipline of information theory had not sunk into the popular imagination. Information, in this context, is not specific; it is, rather, the way the universe is organised. Just as the Greeks saw fire, atoms or water as the ultimate constituents of matter, so we now see information flowing through all things. The universal machine – the computer – produces the universal substance – information.

The contemporary idea of a machine producing an abstraction sprang from the mind of a curious, difficult individual called Charles Babbage, born in London in 1791, who grew up with an intense dislike of numerical imprecision. With heroic pedantry, he once wrote to the poet Tennyson about his lines 'Every moment dies a man, / Every moment one is born.' Babbage pointed out that this would mean the population of the world must be static, but, in fact, it was increasing so the lines, if they were to be accurate, should read, 'Every moment dies a man, / Every moment $1^{1}/_{16}$ is born.'

In 1827 Babbage's father, wife and one of his sons all died. He was in the middle of spending £1,500 of public money, in today's terms about £113,000 (using the retail price index, a great deal more using other measures) on what many regarded as a futile scheme. Babbage was brought to his knees and friends advised him to take a trip round Europe to relax and recuperate, a common remedy in those days. Characteristically, he misunderstood the point of the trip and took the opportunity to do further research on his project. In his absence, he was appointed Lucasian Professor of Mathematics at Cambridge, a post held in different generations by Sir Isaac Newton and Stephen Hawking. Before his return, more voices were raised in criticism of the project, and he also found himself in a bitter financial dispute with Joseph Clement, his assistant. 'I am,' wrote Babbage, 'almost worn out with disgust and annoyance at the whole affair.' Work ended on the project in 1834 and the government officially withdrew in 1842. A total of £23,000 had been spent, £6,000 of which was Babbage's own money.

Pictures of Babbage show a determined but troubled man, perhaps too much the author of his own woes. He was a mathematician of genius and his life's work was to replace the error-strewn mathematical tables of the day with something more exact. The project had been to build a machine that would perform the necessary calculations with absolute accuracy. He called it the Difference Engine because it used the purely arithmetical method of finite differences.

After the failure of Difference Engine 1, Babbage went on to design Difference Engine 2 and the Analytical Engine which was programmable using the punched-card system first devised for the Jacquard loom. Neither of the latter two was ever built in Babbage's lifetime, though Difference Engine 2 was constructed between 1985 and 1991 in celebration of the 200th anniversary of Babbage's birth.

This reconstruction now sits in the mathematics gallery of the Science Museum in London. Made of cast iron, steel and brass, its most intricate and distinctive features are the towers of brass wheels inscribed with numbers. The rest of the machine could be, to the layman, any mighty metal machine of the early industrial age. At one end there is the big crank handle that drives the whole process. Four turns of this crank can produce calculations involving numbers of up to thirty-one digits. It is, says the label, the 'biggest mechanical calculator in the world'. It will remain so. No calculator now needs to be mechanical and only very powerful computers need to be anything like that size. Downstairs in the

Making the Modern World gallery is a severed lump of DE1, poignant in its incompleteness.

When Babbage died in 1871, aged seventy-nine, having never attained the precision of which he dreamed, he did not know that he had inaugurated the information age, the age of abstraction and of intelligent machines. The left side of his brain is preserved in the Hunterian Museum at the Royal College of Surgeons in London. The right side is in the Science Museum.

His Difference Engine is deceptive. It may look like any industrial age machine but it does not behave like one. It does not push anything through water or along rails, it does not weave cloth or spin yarn, it does not pump water. What it does is solve arithmetical problems. This has no effect on the material world whatsoever, or, at least, not until these solutions have been used to change something material. The actual output of Babbage's engine is entirely abstract, a number. It was the beginning of the possibility of a machine that made thought. This confused some of his contemporaries.

'On two occasions,' Babbage drily observed, 'I have been asked, "Pray, Mr Babbage, if you put into the machine wrong figures, will the right answers come out?" I am not able rightly to apprehend the kind of confusion of ideas that could provoke such a question.'

In a curious way, the people who asked that question were over-estimating the machine, thinking it could, somehow, divine the true intentions of the human putting in the figures and correct his input error. Babbage found this absurd. To him, the computer was still a passive tool. But now we have predictive texting and Google routinely works out what we really intended when we mistype. The descendants of his Difference Engine can, indeed, give out the right answers when the wrong numbers are put in.

This moment of transition from material to abstract machine output is the moment when the contemporary – as opposed to the modern – machine age was born. It was the moment when a new model of the human mind came into being, a machine model.

It was also the moment of the birth of the contemporary idea of the immaterial and of the machine. Information was immaterial and the machine was a device that invaded the realm of human competence and aspiration. There is a direct line from DE1 through the call tree to the iPad.

People queue to buy the Difference Engine's descendants because these gadgets have been rendered sacred by their intimacy and the immateriality of what they absorb and produce. They are marketed and designed to insinuate their way into every moment of our lives. Via Twitter they demand that we report our lives and interact with others, via Facebook they ask us to expose ourselves to the world and, via the Cloud, they ask us to disperse ourselves to remote, virtual locations. They connect us to strangers as well as friends. They wrap us in a world of music or video. They make us dependent and vulnerable to their constant demands. Our steady state, as Saul Bellow once said, is distraction.

Technological optimists welcome this as benign connectivity, an expansion of human potential, of life itself. 'Only connect' wrote. E.M. Forster in his novel *Howards End* (1910) and then 'human love will be seen at its height. Live in fragments no longer.' Connection alone is enough to exalt us, to relieve the human predicament. If true, then we live in genuinely exalted times.

But pessimists say: be careful what you wish for. Too much connection may be worse than too little and distraction destroys the mind. Anxieties about this new connected condition emerge all the time. In 2008 Nicholas Carr published an article in *The Atlantic* entitled 'Is Google Making Us Stupid?: What the Internet is Doing to Our Brains'. This later became a book, *The Shallows*. Carr had noticed that he was finding it increasingly difficult to immerse himself in a book or a long article: 'The deep reading that used to come naturally has become a struggle.' Instead he had begun to Google his way through life, scanning and skimming, not pausing to think, to absorb. He felt himself being hollowed out by the process, 'the replacement of complex inner density with a new kind of self – evolving under the pressure of information overload and the technology of the "instantly available"'.

'The important thing,' he wrote, 'is that we now go outside of ourselves to make all the connections that we used to make inside of ourselves.'

The problem with being permanently connected, being enfolded by the comforting immateriality of information, is that you are never not connected. The gadgets are always on and always online. Only in sleep do we escape from their sleeve-plucking demands and, even then, they are storing up sleeve plucks for the morning. There is no down time and down time, surely, is good.

Loren Frank of the University of California at San Francisco argues that freedom from distraction is essential to health. 'Almost certainly, downtime lets the brain go over experiences it's had, solidify them and turn them into permanent long-term memories.'

Research on rats has shown new patterns of activity in the brain when new experiences were encountered. But these patterns were lost – forgotten – unless the rats rested. And, at the University of Michigan, it was found that people learned more efficiently after a quiet walk.

David Meyer is professor of psychology at Michigan. In 1995 his son was killed by a distracted driver who jumped a red light. His speciality is attention – how we focus on one thing rather than another. Attention is crucial to any attempt to understand human consciousness; it may one day tell us how we make the world in our heads. Attention comes naturally to us; attending to what matters is how we survive and define ourselves. The opposite of attention is distraction, an unnatural condition and one that, as Meyer discovered in 1995, can be lethal. Now he is convinced that chronic, long-term distraction is as dangerous as cigarette smoking. In particular, there is the great macho myth of multi-tasking. No human being, he says, can effectively write an email and speak on the phone at the same time.

The most culturally resonant objection came from the environmentalist Bill McKibben. McKibben's hero is Henry David Thoreau who, in the nineteenth century, cut himself off from the mounting distractions of industrialising America, to live in quiet contemplation by Walden Pond in Massachusetts. He was, says McKibben, 'incredibly prescient'. Thoreau's quiet isolation was an extreme attempt to make all time into down time, to live the life of undistracted contemplation.

It feels like a kind of bliss, but it is, admits McKibben, almost unobtainable. He must organise his global-warming campaigns through the internet and thus suffer and react to the beeping pleading of the incoming email. 'I feel that much of my life is ebbing away in the tide of minute by minute distraction ... I'm not certain what the effect on the world will be. But psychologists do say that intense close engagement with things does provide the most human satisfaction.'

McKibben describes himself as 'loving novelty' and yet 'craving depth', the contemporary predicament in a nutshell. He adds: 'The next generation will not grieve because they will not know what they have lost.'

But if they do not grieve, do not know, can anything be said to have been lost? To whom will it have been a loss? The young are usually baffled when confronted by criticisms of their connected world. To them, the eternal on switch is just the world as it is. Mostly, for young and for the old, this is a cause for celebration. If what has been lost was worse than this, what is there to mourn?

From Charles Babbage to Steve Jobs, from the call tree to the iPad, we have constructed the method and the means for the union of man and machine, for the creation of the cyborg. The project is far from complete and both in our loathing of the call tree and our anxieties about distraction and hyper-connectivity we still betray doubts about what Bill Gates called *the* road ahead. But, at the same time, we don't really want to hit zero and subvert the system, we long to be united with these sacred gadgets, the engines of information.

At the Science Museum today the parties of schoolchildren do not make it up to the second floor and the mathematics gallery. Many of them wear the single most brilliant branding coup by Steve Jobs and Jonathan Ive – the white wires of the iPod head-phones. On the ground floor, eager to get past the Victoriana in glass cases to more familiar machines, they do not stop to see the dark metal remains of Babbage's Difference Engine 1, a poignant memorial to a difficult man and the one true precursor of the children's world.

MEN WITHOUT CHESTS

Clive Staples (C. S.) Lewis is an alluring figure – serious but whimsical, joyous but anguished. As a writer he could be inept but also magnificent; as a thinker he was capable of dazzling flashes of insight. He was at his most dazzling in February 1943.

Forty-two years old and a tutor in English language and literature at Magdalen College, Oxford, he was already well known as a literary critic and, having lost his faith in his teenage years and regained it in his early thirties, as a subtle and passionate advocate of Christianity. His later, worldwide fame was to be based on the series of fantasy novels *The Chronicles of Narnia*, written between 1950 and 1956. He died in 1963 but he remains, thanks to *Narnia*, one of the great entertainers of Babbage's children.

In that wartime February, he gave a series of lectures at Durham entitled 'The Abolition of Man'. They were not about the more obviously urgent issues of the moment – Hitler, the threat of Nazism, the confrontation between capitalism and communism or the politics of the post-war world. They were about a primary school book. Yet 'The Abolition of Man' remains one of the most prescient critiques of the second machine age, the information age, our age.

The inspiration of the lectures was bizarre. In all innocence, a publisher had sent Lewis a complimentary copy of a primary school English textbook, presumably hoping for his approval. The publisher did not get it; instead, he was told he was undermining civilisation.

Lewis starts politely, quietly, though ominously. He says he will protect the authors of the book by changing their names and not revealing the title – 'I do not want to pillory two modest practising schoolmasters.' He calls them Gaius and Titius. He then criticises them for what, at first, seems like a rather specialised technicality.

Gaius and Titius tell the story of the poet Coleridge who overheard two tourists at a waterfall; one described the scene as 'sublime', the other said it was 'pretty'. Coleridge endorsed the first and rejected the second with disgust. But the authors use the story in a way that, for Lewis, signals a momentous change in the relationship between the mind and the world.

They claim that the word 'sublime' had nothing to do with the waterfall; it was just an expression of the feelings of the viewer. This, they say, is a general problem with the use of language – 'We appear to be saying something very important about something: and actually we are only saying something about our own feelings.' The implication is that statements of value are merely expressions of emotion and that such expressions are, therefore, unimportant.

Lewis begins to tear this apart by defending the real world, objective existence of values – 'the belief that certain attitudes are really true, and others false' – against the contemporary tendency to turn everything into an expression of individual psychology. There is, for Lewis, such a thing as a natural law, common to all cultures, which he identifies with the Chinese conception of the Tao. Expressions of value are real because they are founded upon this law.

To deny this, to locate values solely in individual psychology, is to reduce the concept of the human, to deprive us of our dignity and diminish the significance of our consciousness. Such a denial will make 'men without chests', people hollowed out by their inability to see values as anything more than self-expression and, therefore, trivial. Such people, Lewis believes, will feel free to conquer nature and, tyrannically, to dictate terms to the future because they will see in the world nothing more than their own preferences and impulses, nothing before which they must bend their knees.

'The final stage is come,' he wrote, 'when Man by eugenics, by pre-natal conditioning, and by an education and propaganda based on a perfect applied psychology, has obtained full control over himself. Human nature will be the last part of Nature to surrender to Man. The battle will then be won. We shall have taken the thread of life out of the hand of Clotho [the Greek fate responsible for spinning the thread of human life] and be henceforth free to make our species whatever we wish it to be. The battle will indeed be won. But who, precisely, will have won it?'

With the best of intentions, the technocrats and futurephiles will have taken away our humanity. 'It is not that they are bad men. They are not men at all. Stepping outside the Tao, they have stepped into the void. Nor are their subjects necessarily unhappy men. They are not men at all: they are artefacts. Man's final conquest has proved to be the abolition of Man.'

Lewis is anticipating the post-human world of the apostles of the Singularity or the transhumanists, though, unlike them, he does not rejoice in the spectacle – no man could be less of a machine-minded geek. More precisely, he is anticipating our own historical moment. Men without chests, men as artefacts, suggest two of the great and, I believe, closely related images of our time – the robot and the celebrity, both creatures that have stepped into the void where values have no objective existence.

* * *

In February 2010, the golfer Tiger Woods gave a fourteen-minute televised speech apologising for his adultery. He later, according to the *National Enquirer*, admitted to his wife of six years, Elin, to 120 affairs. They divorced six months later.

'I have,' Woods said in his speech, 'severely disappointed all of you. I have made you question who I am and how I have done the things I did. I am embarrassed that I have put you in this position. For all that I have done, I am so sorry. I have a lot to atone for.'

The apology was subjected to very close analysis. In particular, people noticed something odd about the tone and manner of his delivery. Wearing an unfashionably loose-fitting jacket and trousers and no tie with a shirt that demanded one, he spoke his words as if in a trance, repeatedly glancing down at his script. Each word emerged as if in isolation from all the others and emotional cues were either missed or over-stated. Some said these were signs of his deep contrition or humiliation, but most concluded the whole thing was false.

The apology came as the climax to a story that began three months earlier with a *National Enquirer* exposé of one of Woods' affairs. A couple of days later, he was involved in a car accident outside his home after which he refused to speak to the police. He withdrew from immediate golf commitments. Numerous women suddenly started coming forward with stories of affairs. Woods made a statement offering an apology, but corporate sponsors – Gillette, Tag Heuer, AT&T and Accenture – began to withdraw support.

That month the University of California Davis Graduate School of Management produced a study suggesting that shareholders in sponsor companies had collectively lost between $5 and $12 billion as a result of the scandal. The study covered the thirteen trading days between 27 November, the date of the car crash, and 17 December, when Woods announced he was leaving golf indefinitely.

'Does celebrity sponsorship have any impact on a firm's bottom line?' asked Victor Stango, one of the study's authors. 'Our analysis makes clear that while having a celebrity of Tiger Woods' stature as an endorser has an undeniable upside, the downside risk is substantial too.'

The risk being, of course, that superstar sportsmen might not be quite what they seem. But Woods had seemed perfect. In June 2008, over a year before the scandal broke, David Brooks had written incisively about the spectacle of Woods on the golf course in the *New York Times*: 'In a period that has brought us instant messaging, multitasking, wireless distractions and attention deficit disorder, Woods has become the exemplar of mental discipline.'

Brooks quoted David Owen writing in *Men's Vogue* who had explicitly made a robotic connection. 'Woods's concentration often seems to be made of the same stuff as the liquid-metal cyborg in *Terminator* 2: If you break it, it reforms.'

Brooks clearly admired Woods' mental strength but he was ambivalent – 'You can like this model or not' – partly because of the way this strength had been bought and sold. Woods had become 'the beau ideal for golf-loving corporate America.'

But there was a more profound unease about the 'frozen gaze' of the golfer. 'Woods seems able to mute the chatter that normal people have in their heads and build a tunnel of focused attention.'

Woods' behaviour in his personal life was widely condemned. In December 2009, Jan Moir in the UK's *Daily Mail* responded to his initial apology which had been posted on his website:

He pleaded to establish 'an important and deep principle' which was the 'right to some simple, human measure of privacy'.

Perhaps he should have thought of that before 'sex-texting' random smut buckets and hooking up with hot babes while on golf tournaments.

Apology and condemnation are, on the face of it, odd. Tiger Woods is revered for his golf; why should his private life matter? But it does matter because Tiger Woods is not just a golfer, he is a celebrity. It is almost certainly true to say that most people who followed the scandal were not remotely interested in golf; they were simply interested in the story of a famous person behaving badly. Golf was no more than the contingent cause of his availability as a mechanism for moral outrage.

Justifying outrage in such terms would be impossible were it not for the invention of the category of 'role model'. People must care about the private life of the famous precisely because they were famous and, therefore, potentially imitable by the young and the impressionable.

John Terry was the England football captain before he was laid low by a sex scandal, also in February 2010. 'The role of the role model,' wrote Jan Moir again about Terry, 'in society is sometimes exaggerated, but not in football. It is a sport that obsesses millions. Children and teenagers in particular are spellbound by football stars.

'Their habits are studied, admired, copied. Children revere them ... The argument that John Terry and big football stars like him do not set an important example is nonsense.'

Previously the media did not report the misdeeds of role models. In the days when being gay was a misdeed, Rock Hudson was protected from exposure as a model of American manhood. The British royal family enjoyed years – centuries – of deferential discretion until tabloid competition determined that they too must be available for moral outrage. Now the primary function of celebrities is to bear the burden of high moral obligation and then to drop it as spectacularly as possible.

Sports stars, in particular, had to bear another burden – sponsorship. Tiger Woods was wreathed in high-price sponsors. His annual earnings were said to be $90.5 million, though *Forbes* magazine estimated $105 million. Golf only contributed about 20 per cent of that figure. Playing well, therefore, was a necessary but only a small part of his professional repertoire. His primary task was to be the sort of person companies could comfortably attach to their products. Money – the profits of tabloids and sponsors – insisted that Tiger Woods could not just be a golfer; he must also be a moral puppet.

Or, more precisely, robot. His apology speech was, indeed, robotic. After the revelation of his all too human sex life, we might have expected that this professional façade would crack open to reveal a real human being. But, in fact, nothing cracked. What was striking about his speech was not that it was contrite, emotional, awkward, false, frank, calculating, targeted, shabby, self-interested, generous, nor even that it was plainly written by lawyers and PRs. No: what was really strange was that it was an algorithm, a set of instructions, Tiger's software for the day.

Our first robots are not made of metal, they are made of flesh. The men without chests, foreseen by C. S. Lewis, have arrived. Celebrities are a dry run for the fully machine future, our best robots yet. They are, in fact, what roboticists would call 'affective robots'; they display emotions with which humans can identify. Whether they actually experience these emotions or not is beside the point.

'Finally,' concluded Woods' speech, 'there are many people in this room and there are many people at home who believed in me. Today, I want to ask for your help. I ask you to find room in your hearts to one day believe in me again. Thank you.'

In the audience was Tiger's mother, Kultida, seated between Kathy Battaglia of Tiger Woods Enterprises and Amy Reynolds of Nike. After the speech Kultida hugged her son.

It has taken more than fifty years to arrive at such a refined version of the robot celebrity. The celebrity as a distinct sub-category of the human was tentatively formulated in 1962 by the American historian Daniel Boorstin who coined the phrase 'famous for being famous' as a way of capturing the way in which modern media detached fame from actual achievement and made it into an autonomous category so that merely being famous became a self-contained aspiration. Now it is taken for granted that to be a celebrity requires no prior qualification or achievement.

Some, in fairness, see the absurdity and the danger in this, among them Cheryl Cole, a British A-list celebrity because of her career as a pop singer, her failed marriage to a footballer and her role as a judge on the talent show *The X Factor*. Asked if she was embarrassed about the newspaper exposé of her ex-husband's infidelities, she gave a highly significant response.

'Of course. I'm a human. I'm still a person, you know. I know to a lot of people the headlines and the stories they read are like some sick entertainment or soap opera, but it's my life and I'm really dealing with it, and it's really happening. It's my real life. Of course I was embarrassed.'

Her words capture the twofold anxiety that being a celebrity threatens to turn her into something non-human and that people may not be able to grasp that there are real events in her life involving real suffering. Or perhaps they think that being famous softens all blows.

The internet intensifies this process. Cole was asked if she could live a normal life.

'I can act normally, and I can go out, but people around you don't act normally. If it was normal, I'd go out and have my dinner and nobody would be video-phoning me eating my dinner.'

You cannot just be spotted in a restaurant, not just even photographed; you can be filmed on a smartphone and immediately the spectacle of your dinner is distributed around the world.

Actors are especially vulnerable to being consigned to the subcategory of celebrity. Of the hundreds of interviews I have conducted over the years, actors are invariably the most difficult. In my interviews, I am not particularly interested in people's private lives – unless there is some real relevance – because, on the whole, all private lives tend to be variations of the same rather narrow collection of themes, themes that are better covered by novelists or gossips and are certainly unlikely to be exposed in any depth within the time restraints of the average interview. Unless the interviewee is solely famous for being famous, the reason they are being interviewed is their work – usually, in my case, their art – so that is the focus of my interest.

This is why actors are so difficult. The majority of them are utterly incoherent on the subject of what they actually do. This is a tribute more than a criticism – good actors are usually so deeply buried in their art that they find it hard to explain. Also having nothing to say *for themselves* is the purpose of actors. Their trade is to become others and, in doing so, they become extensions or versions of ourselves and the people we encounter. But actorish aphasia does make life difficult for the journalist and for the retinue of publicists that attend the most famous actors. The usual solution is to resort to the private life, specifically to the love life, or to a cause, usually environmental, which the actor has embraced in what often seems like a rather pitiable attempt to escape the conversational confines of their inexplicable craft.

Exceptions prove the rule. Dinner with Helen Mirren, for example, consisted of a running battle with her refusal to play any kind of interview game.

'What do you do when you're not working?'

'Not working? What the fuck? I don't know, read the Sunday papers, watch a bit of TV. What the fuck do you do when you're not working? I don't have a hobby, I don't make things out of matchsticks.' She was not, I wrote, anything like 'the flashy, pumped-up, dead-eyed robots who dominate the movies'.

Less successful was Ricky Gervais's determination to turn the entire interview into a performance. He was also trying to escape the suffocating prison of the conventions of contemporary fame.

But, for the most part, celebrities in general and actors in particular have become men without chests, people in whom the inner life seems to have been destroyed by the outer. They are victims of the complicated machinery of fame. After many failures, the science of robotics has its first convincing humanoids.

Robot, the word, was suggested to the playwright Karel Čapek by his brother Josef. *Robota* means, in Czech, work or drudgery and Čapek's play of 1920, *R.U.R.* (*Russum's Universal Robots*), was set in a factory that made humanoid machines. Robotics was coined as the title of the science by Isaac Asimov. Both 'robot' and 'robotics' have an appropriately hard, mechanical sound which probably explains the success of the words in English.

The factual, as opposed to fictional, development of the idea in the post-war period represents one of the most important and profound intellectual projects of our time. Presidents and prime ministers have come and gone in the last seven decades; none has been as important to the creation of the contemporary world and to our future as the scientists and thinkers who pursued cybernetics, artificial intelligence, information theory and robotics.

It all began in 1942, the year before C. S. Lewis delivered his 'Abolition of Man' lectures. Frank Fremont-Smith, an executive with the Josiah Macy Jr Foundation, set up a meeting at the Beekman Tower Hotel in Manhattan's Midtown East. Attendees included neurophysiologist Warren McCulloch, physiologist Arturo Rosenblueth, anthropologist Margaret Mead, anthropologist, linguist and Mead's then husband Gregory Bateson and psychiatrist Lawrence Kubie. The intention was to discuss 'physiological mechanisms underlying the phenomena of conditioned reflexes and hypnosis as related to the problem of cerebral inhibition'. It is generally known as the Cerebral Inhibition Meeting.

In fact, the true subject of the meeting – and the reason it is remembered at all – was defined by Rosenblueth's startling contribution. He was 'a burly, vigorous man of middle height, quick in his actions and speech,' according to the mathematician Norbert Wiener, Rosenblueth's collaborator at Harvard. His speech concerned the new science they had jointly created.

He told tales of torpedoes that guided themselves to their targets by seeking out the magnetic attraction of a hull or the sound of a propeller, of plants that seek the light, of bodies that control appetites and temperature. Rosenblueth was describing what he called 'circular causality'. These machines and organisms constantly check back to assess their progress towards their goal; they use 'feedback loops'.

All these things had purposes, they exhibited 'goal-directed' behaviour – the torpedo sought the ship, the plant the light and our bodies wanted to maintain the right temperature. This may seem trivial or obvious but it was, as Rosenblueth acknowledged, heresy, which is why both he and others at the meeting were convinced they were on the threshold of a scientific revolution.

Activities directed towards a goal are called, in philosophy, teleological. Teleology is the belief that nature itself is goal-directed, that it is moving towards something, whether that is the kingdom of God or the just society. Most scientists have long rejected teleology in nature because it implies a material impossibility – that the future can affect the past. A hard-line teleologist might argue that the ape was purposefully moving towards evolving into a human or that the first replicating molecule was busily building the road to heaven. But this, to the materialist, is mysticism.

On the other hand, teleology is all around us. A mouth and teeth assume a future which involves food, just as the fire in a steam engine assumes a future involving movement. Without this form of teleology, the world becomes incomprehensible. The new science born at that conference at the Beekman would reinstate teleology as a fundamental tool for understanding and changing the world.

Feedback loops are the heart of the matter. They give purpose to a machine or an organism. The concept is very straightforward: what happens next is determined by what just happened.

'The nervous system and the automatic machine,' wrote Norbert Wiener, 'are fundamentally alike in that they are devices which make decisions on the basis of decisions they have made in the past.'

You step into the road, see a car coming and step back on to the pavement. The step back is determined by the step forward. Or a thermostat turns on the heating because the temperature is below the set level, but then turns it off when it reaches or exceeds that level. It is continually checking on the effects of its preceding actions, just as you checked on the road. The important point – the imagination transforming point – is that this mechanism is the same in people and machines.

When the war ended, Bateson and McCulloch, excited by what they had heard, pressed Fremont-Smith to organise a follow-up meeting. As a result, between March 1946 and April 1953, a series of conferences were held in New York and Princeton. They were again organised by the Macy Foundation, a philanthropic body devoted to improving healthcare. But the Macy conferences went much further than improving health. They were designed to create the basis of a general science of the human mind.

As a picture of, largely, American intellectual and scientific life in the immediate aftermath of the Second World War, the conferences were unparalleled. As a foreshadowing of the primary preoccupations and technologies of the post-war world, they were unique.

Psychiatry, anthropology, linguistics, philosophy, computer science, mathematics, information theory, electronic engineering, physics and biology were all represented. The core group included Mead, mathematicians John von Neumann and Wiener, Bateson and McCulloch; guests included sociologist Talcott Parsons, bio-physicist Max Delbruck and, most important of all, electronic engineer, mathematician and creator of information theory Claude Shannon.

Computer science and biology dominated the discussions. Both were at a critical juncture. As a result of the war effort, digital computers – the Colossus in Britain and ENIAC in America – had been built and, thanks to Alan Turing, Shannon and von Neumann, the theoretical foundations of its development had been laid. Meanwhile, in biology, the race to decipher the DNA molecule was on, finally ending when Watson and Crick won with their double helix model. Their discovery was made in February 1953 and published in April, just as the Macy conferences were ending. It seemed to show that life itself was digital, like a computer.

The world was to be digital; it could be broken down into numerical units. The now superseded analog technology is continuous, it compares reality to the reaction of chemicals on a celluloid film, the infinite variations of an electrical current or the surface configuration of a vinyl record. But, with digital, the world is converted into a form readable by digital computer; it is broken down into a series of discrete units.

The conferences were confusing. There were ten in all and the first five have left no proper record except in attendees' notes. But it was clear that they were chaotic and rancorous. 'The first five meetings were intolerable,' remembered Warren McCulluch, 'the smoke, the noise, the smell of battle were not printable.'

'Nothing,' he went on, 'I have ever lived through ... has ever been like ... those meetings ... You never have heard adult human beings, of such academic stature, use such language to attack each other. I have seen member after member depart in tears.'

The conferences remain a blurred image in scientific history, but the big picture is in full focus. What was being created was the science of cybernetics, the science that destroys the barrier between humans and machines. The word derives from the Greek word for governor, helmsman or rudder. It seems to have been first used by Plato in the context of government. It took on its modern usage with the publication of Norbert Wiener's *Cybernetics: or Control and Communication in the Animal and the Machine* in 1948.

Cybernetics is easily defined – it is the science of communication and regulation mechanisms – but less easily understood. Why should there be such a science? The reason is the fragmentation of the other sciences and technologies.

Biologists may look at evolution, and physicists at relativity and quantum theory. Engineers may design machines, and surgeons fix the human body. But who looks at what links all these things? The answer was the cyberneticists, the scientists who examined the underlying mechanisms and who were, therefore, meta-scientists. Only they, by their command of communication and control mechanisms, could aspire to understand nature as a whole.

Ironically, in the ensuing years cybernetics as a distinct discipline was to be subjected to the very fragmentation its founders were trying to cure. It fractured into artificial intelligence research, affect control theory, computer science, robotics, control systems, cyborg research and countless other sub-disciplines. Nevertheless, those foundational ideas first outlined by Rosenblueth at the Beekman Tower Hotel have infiltrated almost every aspect of post-war technology. At their heart is the ever closer connection between man and machine.

But is this a good development? There were three ways of looking at the ethics of the destruction of the human–machine boundary. First, it may morally improve humanity, rescuing us from our fallen state.

'I don't particularly like people,' said Warren McCulloch. 'Never have. Man to my mind is about the nastiest, most destructive of all the animals. I don't see any reason, if he can evolve machines that can have more fun than he himself can, why they shouldn't take over, enslave us, quite happily. They might have a lot more fun, invent better games than we ever did.'

Or it may be a very good thing because it would protect our humanity. In a crucial passage in *Cybernetics: or Control and Communication in the Animal and the Machine*, Wiener, now generally regarded as the father of cybernetics, outlines the humanist, liberal, problem-solving view of this new science:

It has long been clear to me that the modern ultra-rapid computing machine was in principle an ideal central nervous system to an apparatus for automatic control; and that its input and output need not be in the form of numbers or diagrams but might very well be, respectively, the readings of artificial sense organs, such as photoelectric cells or thermometers, and the performance of motors or solenoids ... Long before Nagasaki and the public awareness of the atomic bomb, it had occurred to me that we were here in the presence of another social potentiality of unheard-of importance for good and for evil.

Perhaps I may clarify the historical background of the present situation if I say that the first industrial revolution, the revolution of the 'dark satanic mills', was the devaluation of the human arm by the competition of machinery ... The modern industrial revolution is similarly bound to devalue the human brain ... The answer, of course, is to have a society based on human values other than buying and selling. To arrive at this society, we need a good deal of planning and a good deal of struggle.

Or – a variation of this view – cybernetics could save us from final destruction. We could discover new ways of, at least for a while, fending off the ever-increasing wave of disorder – entropy – that must, the Second Law of Thermodynamics tells us, one day engulf us all.

Of all the Macy attendees, one stood out as a giant of the contemporary world and its preoccupations. Born in 1916 (he died in 2001), Claude Shannon has a strong claim to be the most influential thinker of the twentieth century. He could also juggle three balls while riding a unicycle. He colluded with Ed Thorp on a method of beating the house at roulette. Thorp – who went on to become the greatest of all quants (quantitative analysts who try to control financial market risks with sophisticated mathematics) – had already worked out a way of beating the house at blackjack, a method documented in his book *Beat the Dealer* (1962).

Shannon also invented the Ultimate Machine. This was a box with one on-off switch. Turn the switch on and a lid opens, a hand appears, turns the switch off and retreats back into the box. The sole function of this machine is to keep itself turned off. The science fiction writer Arthur C. Clarke described the machine as 'unspeakably sinister'. It is now made in various forms by enthusiasts and is generally know as the Leave Me Alone Box. This is more accurate than Ultimate Machine because, strictly speaking, it isn't ultimate. If it were truly ultimate, it would work only once. The off mode would be final.

Entropy – randomness – was as important to Shannon as it was to the cyberneticists. But, to him, it was a positive force. It was the background noise against which information could be detected and Shannon's entire career was about information.

He was both a mathematician and an engineer, a 'tinkerer', a practical as well as an intellectual adept. His 1937 master's thesis at MIT, 'A Symbolic Analysis of Relay and Switching Circuits', is the foundation of the digital world. He showed how the electromechanical relays then used in telephone systems could be massively improved by using the algebra of George Boole, a nineteenth-century mathematician. Boole had used algebra to create a logical system aimed at arriving at a system of truth values. This was based on a binary code of zeros and ones.

In telephony a binary code for voice signals would solve the most serious outstanding problem. Analog telephony is subject to entropy. The voice signal decays as it is transmitted over long distances and noise – entropy – takes over. What was needed was a way of maintaining the clear distinction between noise and signal. The answer was to code the signal digitally so that it preserved a clear pattern. Central to this was error correction – by adding more code, Shannon could make signals self-correcting.

You can see this most clearly in the difference between a vinyl record and a CD. Scratch an LP and the scratch will be played as noise alongside the signal. Scratch a CD and error correction means you will still hear the signal alone.

Shannon went on to specify binary code – a series of ones and zeros or ons and offs – as the logical language for digital computers. The Ultimate Machine may be seen as a kind of bleak commentary on this idea as it makes a futile game out of the simple facts of on-ness and offness.

Binary code implanted in the modern imagination the belief that everything can, ultimately, be broken down into a primitively simple and irreducible pair of atomic units that specified only that there was something or there was nothing.

The effects of this idea were momentous and, strangely, in view of its roots in mathematics and computer science, it ushered in a new anti-materialism in the form of the cult of information. The futurist Alvin Toffler was to become its most celebrated prophet. He argued that human society had been subject to three waves. The first, 10,000 years ago, was the overthrow of mobile hunter-gatherer cultures by static, agrarian ones. The second, 250 years ago, replaced agrarianism by the industrial culture of mass consumption and production. The third is the post-industrial society that began to emerge in the post-war era, the era of computers, mind and information. Toffler called our era the Information Age and it is that word 'information' which marks Claude Shannon out as the one true surfer of the new wave.

Shannon defined information for the post-war world in a paper written in 1948 entitled 'A Mathematical Theory of Communication'. 'The fundamental problem of communication is that of reproducing at one point either exactly or approximately a message selected at another point. Frequently the messages have meaning; that is they refer to or are correlated according to some system with certain physical or conceptual entities. These semantic aspects of communication are irrelevant to the engineering problem. The significant aspect is that the actual message is one selected from a set of possible messages. The system must be designed to operate for each possible selection, not just the one which will actually be chosen since this is unknown at the time of design.'

The idea has been familiarised by the way we live now, but it still sounds shocking and strange. Information, for Shannon, was content free, the message was always blank, defined only by its recognisability as a pattern distinct from chaos. Like the binary code, it is a Platonic perfection. And, indeed, it was in that paper that Shannon christened the basic atomic unit of computing – the binary digit. He called it the 'bit'.

The problem with content is that it implied that something must be said about both the sender and the receiver. Did they, for example, both speak English? Or even were they happy or sad? This would have immediately destroyed the possibility of information theory becoming anything like a hard science. It would need to take in notoriously soft sciences such as psychology, sociology and linguistics. This was not good enough for Shannon and certainly not good enough for the coming information age. Strict quantification was needed if the new machines were ever to get going.

It is possible that Alan Turing inspired Shannon's work in information theory. During the war, when Turing was working on the breaking of German codes, he visited Shannon at Bell Laboratories in order to talk about speech encoding. They worked together on an encrypted phone that would allow Churchill to hold secure transatlantic conversations with Roosevelt.

They had much in common, most obviously a fascination with the idea of intelligent machines. But they also shared a high rationalist, as opposed to empirical, view of the world. Shannon's definition of information as content free was the correlative of Turing's Test in which machine intelligence could be established through entirely disembodied means.

So, in the immediate post-war period as technologies developed to defeat the enemy were reapplied to the possibilities of a new science and a new social order, the foundations were laid of the world in which we now live, a world in which information as abstraction is idolised, machines are conceived as ever more intimate and humans as ever more like machines.

The empty message, cybernetics and the binary computer came together to produce one of the most pervasive and persistent ideas of the post-war world. The idea springs naturally from the extreme Platonism of the binary code. If everything can ultimately be specified by zeros and ones, then so can humans. And if the binary code is the language of computers, then we can be replicated on a computer. The human individual is, like everything else, just another pattern. 'We are,' wrote Norbert Wiener, 'but whirlpools in a river of ever-flowing water. We are not stuff that abides, but patterns that perpetuate themselves. A pattern is a message, and may be transmitted as a message.'

Or a pattern may be uploaded. Human consciousness may be no more than strings of zeros and ones. Hans Moravec of the Robotics Institute of Carnegie Mellon University imagines a robot surgeon sucking out a human brain, reading all the information it contains and then uploading it on to a computer. The person awakens to find himself just as he was and unaffected by the fact that he is now inside a machine.

If we can be uploaded, then, plainly, this is a glimpse of immortality. When uploaded, we can be backed up indefinitely. But we would remain disembodied inside a computer. At first, assuming we were a perfect copy, this would be no burden as we would feel as embodied as we did when we were biological. But this would decay over time unless a suitably stimulating virtual environment was created. This is exactly what happens in the film *The Matrix* (1999), but that is about a future that is a machine-made hell in which the machines see humans as no more than a fuel source. They are kept passive by the creation of a 'real' world that is, in fact, a computer simulation.

Robotics could provide bodies for our uploaded minds, but it would be far better if these superior bodies had superior minds. This is exactly how Moravec sees the future, as a place where biological evolution ends and machine evolution begins. 'Unleashed from the plodding pace of biological evolution, the children of our minds will be free to grow to confront immense and fundamental challenges in the larger universe. We humans will benefit for a time from their labors, but sooner or later, like natural children, they will seek their own fortunes while we, their aged parents, silently fade away. Very little need be lost in this passing of the torch – it will be in our artificial offspring's power, and to their benefit, to remember almost everything about us, even, perhaps, the detailed workings of individual human minds.'

The upload may also be a useful by-product of the arrival of the Singularity in, according to Ray Kurzweil's estimate, 2045. At this point, you will remember, our technological progress converges on the building of our last machine, a conscious computer capable of booting itself into ever higher levels of intelligence. This could show us how to become medically immortal or it could upload us all in one rapturous instant. Or the superior machine intelligence could just leave us behind.

But the real interest of the upload lies in its role as an expression of the accepted truth of information theory and the validity of the Turing Test when applied to human consciousness. The idea that we could be uploaded intact into a computer is based on two assumptions: that binary code is a substrate so robust that it can replicate the human mind and that the disembodied responses of the successfully completed Turing Test are, indeed, conclusive evidence of mind and, by extension, of interiority.

As the philosopher Colin McGinn has pointed out, these are pretty big assumptions. The Turing Test, he argues, is seriously flawed and an expression of a now discredited view of the mind – that it is no more than outward behaviour. 'First,' he writes, 'it is just an application of the doctrine of behaviorism, the view that minds reduce to bodily motions; and behaviorism has long since been abandoned, even by psychologists. Behavior is just the evidence for mind in others, not its very nature. This is why you can act as if you are in pain and not really be in pain – you are just pretending.'

McGinn also points out that we have absolutely no idea how we are conscious. Consciousness seems to reside in the brain but we don't know where. We cannot begin to consider an upload into a machine until we know how to make that machine capable of consciousness and there is no prospect of us ever being able to do this. This raises the nightmarish possibility that the upload may make you an immortal but nonsentient being, a proficient answering machine but one devoid of all interiority – like certain interviewees I have known.

A BBC reporter has just said of a departing politician, 'He spoke human.' The ability to speak human is now seen as unusual in politics. The world of the internet, Twitter, blogs, carefully calibrated initiatives and rebuttals, ranks of 'special advisers' and the need to control and monitor every encounter has made people aware of the difference between human speech and that of the 'machine' politicians or, as C. S. Lewis christened them, the 'men without chests'.

Machines are moving towards people and, especially in the public realm, people are moving towards machines. It is happening because of the near universal – though, perhaps, unconscious – acceptance of the cybernetic, information theoretical, robotic account of the world. But is there really such an account of the *human* world? Surely this account describes something quite different.

'And all the time,' C. S. Lewis wrote, '– such is the tragi-comedy of our situation – we continue to clamour for those very qualities we are rendering impossible. You can hardly open a periodical without coming across the statement that what our civilization needs is more "drive", or dynamism, or self-sacrifice, or "creativity". In a sort of ghastly simplicity we remove the organ and demand the function. We make men without chests and expect of them virtue and enterprise. We laugh at honour and are shocked to find traitors in our midst. We castrate and bid the geldings be fruitful.'

At almost exactly the same time that Lewis was writing these words, the cyberneticians were, indeed, attempting to abolish man by driving him into the arms of the machines. They were, as Lewis feared they would, dictating terms to the future.

PIMP MY ULTRASOUND

Eli Pariser noticed he was not hearing from his conservative friends on Facebook. He is on the left, an activist and president of MoveOn.org, a liberal advocacy group, but he likes to listen to the other side and he has plenty of friends on the right. But, suddenly, what they were saying on Facebook was not appearing on his front page. He also noticed something else. When he put the search term 'Egypt' into Google, the lead results were dominated by the revolution that overthrew Hosni Mubarak. When a friend did the same, the lead results were all about the joys of travel in Egypt.

For a long time I assumed, like everybody else, that the internet was a window and, through it, I could look at whatever I liked. But what Pariser discovered is that the back of this window is now being silvered over. It is being turned into a mirror.

The change was announced – very quietly – on 4 December 2009. From that date Google would use fifty-seven data points – your search terms, your location, your surfing history – to work out what results you should receive. The change was momentous. Previously Google's results were as neutral as an encyclopaedia. If you and I both look up Egypt in an encyclopaedia, we both see the same thing. So it was with Google. But increasingly, after 12 December, we would start seeing different things.

'We're still happy to be in search,' says Eric Schmidt, of Google and the most quotably chilling figure in Silicon Valley, 'believe me. But one idea is that more and more searches are done on your behalf without you needing to type. I actually think most people don't want Google to answer their questions. They want Google to tell them what they should be doing next.'

Now, for good commercial reasons, the internet is becoming personal, horrifyingly so. The site dictionary.com, which I use all the time, installs 223 cookies and beacons on my computer so that other companies can know what I searched for and tailor their ads and marketing to my interests.

Pariser argues we are entering the 'filter bubble'. On the internet, we will discover not the real world, but a filtered version designed to profit from our inclinations and impulses. Since, according to Schmidt, children will only have two states, 'asleep or online', this means that the filtered world will become *the* world, a state of affairs Pariser compares to the machine future evoked by Dostoyevsky in *Notes from the Underground*, a future 'in which everything will be so clearly calculated and explained that there will be no more incidents or adventures in the world'.

In truth, the internet was already well on the way to becoming one big filter bubble long before 4 December. It was and remains dominated by the male hacker and geek culture of the Eighties and Nineties. A quarter of all websites are said to be pornographic, a quarter of all search terms are porn-related and 35 per cent of all downloads are of porn. There is also an excess of trivial, zany, adolescent humour. Bill Gates may not have foreseen the imminence of the internet in August 1994, but, even if he had, could he possibly have foreseen the popularity of lolcats (laugh out loud cats), pictures of cats doing or supposedly saying funny things, or Will it Blend?, a promotion site for Blendtec kitchen appliances on which various improbable objects – an iPhone, a video camera – are tossed into a blender, or the endless embarrassments of home videos that somehow went viral like that of the Star Wars Kid, a boy using a golf ball retriever as the light sabre of a Jedi knight? These may not be the important things on the net, but they are certainly big things.

Seventeen years after Chris Peters of Microsoft confessed the company was not looking for 'well-roundedness in people', much of the internet remains the dream of unrounded freaks. Moreover, these strange, juvenile things feel like inventions of the internet. This technology is not, as we like to think, neutral, a playground designed by its users. Pariser shows that to an increasing extent it is designed by big companies, but also it is designed by the form of the technology itself. It is set up to do all the strange things it does.

There is an iPhone app – application – called Pimp My Ultrasound. This allows you to add comic details to an ultrasound picture of your baby in the womb – sunglasses, baseball cap, credit card, ribbons in the hair, guitar, speech bubbles. It was developed by an internet marketing company in Atlanta called 2 Wise Guys. The company describes itself as 'just a small team of cool, hip people'. Their strategy is not that of the 'shotgun approach' of old media that simply hopes customers will find their advertisements.

As marketing, Pimp My Ultrasound makes perfect sense. Diagnostic sonography uses sound at frequencies too high for human hearing. It is, like fMRI, a way of picturing the interior of the body. It is now familiar in ordinary conversation – where it is known as ultrasound – because of its ability to provide images of the human foetus in the womb. These prenatal pictures are new and imaginatively potent. It is not that they tell us anything we did not know before – although, to the trained eye, they do – rather, it is the fact that they show us what we knew with an entirely novel graphic intensity.

Ultrasound scans have taken on some of the attributes of ordinary, amateur photography. They are used as personal, intimate, family mementos, celebrations of a rite of passage. Like photographs, they are possessed of a sacred, poignant feeling, as if they have overcome forgetfulness and death by fixing a moment in time. Even the blurred ambiguity of the images gives them something of the nostalgic charm of old photographs. But, just as photographs can now be enhanced and altered at will, so ultra-sounds can be improved by digital technology.

The additions to a scan by the Pimp My Ultrasound app at once remove the image from the sacred to the profane world of spending and ordinary fun. The potency is tamed, domesticated and controlled. The relative innocence of the science of sono-graphy, and of the foetus itself, is replaced by the familiar, knowing forms of consumption and of celebrity. Still in the womb, the child has already acquired the trappings of a new type of fame. And as analyst of the internet age Clay Shirky remarks, 'Once a medium gets past a certain size, fame is a forced move.' The medium, in this case, is the machinery that makes the app and disseminates the results, forcing you to make your baby famous.

This kind of thing troubles Jaron Lanier. Lanier, Silicon Valley player, computer scientist and musician, a master of many strange instruments, was named by *Time* magazine in 2010 as one of the world's 100 most influential people. He is also a contemporary C. S. Lewis, concerned about the rise of the men without chests.

Lanier's rise to the higher ranks of the technocracy began in the early 1980s when he first came across the 'data glove', a device for direct human–machine interaction. He subsequently became one of the pioneers of virtual reality, the simulation of a real-world environment within a computer. Latterly, he has been one of the creator's of Microsoft's Kinect game system.

When we spoke, what I found most striking – and significant – about him was his self-deprecating awareness of the psychological basis of his fascination with virtual reality. The key psychological component was his memory of his mother. She was a Holocaust survivor who died when he was nine. 'I was not ready for my mother to die,' he said, 'we were just super-close. I mean I was an only child, she was a concentration camp survivor, we are living in this funny place where I basically didn't have any friends or social reference at that age and – aw, man – I somehow was just so overcome with the experience of grief that the subjective life was far more intense than the perception of objective reality.'

He became, he says, hyper-sensitive to the most subjective elements of experience. This overwhelmed him, making social relations all but impossible.

'That whole world of the subjective was so separated from the world of commonality. The distinction was just too severe. I became obsessed with the idea that maybe some day the distinction didn't have to be so severe that I wouldn't have to lose so much in order to connect with the other people. So the notion was to build a machine that would, if you like, be able to convey some more intense or more comprehensive or more complete form of art that could create a channel between this intense subjectivity and the world of commonality.'

The trauma of his mother's death had flung Lanier into the half-life of the geek; he sought salvation in the technology of escape.

He was born in New York in 1960 but the family ended up in New Mexico. His performance at school collapsed after the death of his mother, as did the family fortunes. Somehow, he found his way into art college in New York State but then left and, aged seventeen, he tried to make a living as a musician in New York. Finally, at eighteen, he returned to New Mexico, convinced his life was over.

Then, in pursuit of a girl who had dumped him, he drove to Los Angeles and, finally, seeking work, he arrived in Silicon Valley. 'I went to this employment office and they said I could get this job or that job and the salaries just floored me. It was like so much money that I was just dizzy.'

He rose higher and higher in the ranks of internet thinkers and scientists but then, in 2006, his doubts got the better of him and he became a Silicon Valley apostate. He published an article in *Edge* magazine entitled 'Digital Maoism: The Hazards of the New Online Collectivism' in which he described 'a resurgence of the idea that the collective is all-wise, that it is desirable to have influence concentrated in a bottleneck that can channel the collective with the most verity and force'.

He went on: 'This is different from representative democracy, or meritocracy. This idea has had dreadful consequences when thrust upon us from the extreme Right or the extreme Left in various historical periods. The fact that it's now being re-introduced today by prominent technologists and futurists, people who in many cases I know and like, doesn't make it any less dangerous.'

In accusing the digerati of neo-Maoism, Lanier was taking on a specific aspect of their creed. This was the cult of the crowd which, along with the filter bubble, defines the net as we know it.

Belief in the virtues of the crowd began with a story told of a cousin of Charles Darwin, a statistician named Francis Galton. In old age, Galton watched a competition to guess the weight of an ox at an agricultural show. Eight hundred people took part; some were experts but most were amateurs. After the contest, Galton collected all the tickets and worked out the average guess. It was 1,197 pounds; the ox weighed 1,198 pounds. The crowd as a whole had proved more expert than the smartest individual farmer.

As brilliantly resurrected and popularised for the internet generation by the writer James Surowiecki, this story, and many similar ones, established the idea of the crowd as cleverer than experts but also as more decent than individuals or small groups. This is, in short, a moral as well as a practical insight and it seems unarguable.

But the case weakens when it becomes clear that the crowd in question has to be carefully specified by external rules. It cannot be a simple collective. Certain types of communication and cooperation between crowd members will destroy the benign dynamic because individuals would become more conscious of the opinions of others and would feel pressure to conform. The crowd only works when it is a set of individuals thinking for themselves. So somebody has to manage the crowd to ensure that it does not separate into factions.

In spite of this caveat, Surowiecki regards crowd wisdom as a hard-wired aspect of human nature. 'With most things,' he writes, 'the average is mediocrity. With decision-making, it's often excellence. You could say it's as if we're programmed to be collectively smart.'

The great internet example of the wisdom of crowds – and of the necessity of external regulation – is Wikipedia, launched in 2001. This is an online encyclopaedia whose sixteen million articles are written by everybody who cares to contribute, expert or not. Because people are also allowed to edit articles, the crowd imposes a self-correcting mechanism into the system. Wikipedia does, indeed, achieve high levels of accuracy and topicality. But, as with Surowiecki's ideal crowds, there needs to be a level of external control by an elite editing group to prevent factionalism and simple abuse.

This raises questions about the exact status of the crowd – can it be autonomously effective or will it always need regulating? If the latter, can it really be said to be a fundamental aspect of human nature? Perhaps the need to regulate is more fundamental.

Clay Shirky is a writer on the social impact of the internet who is destined to be remembered as one of the technology's earliest and most effective spokesmen. Like Surowiecki, Shirky bases his case on the power of crowds. He speaks of the 'epochal' transfer of powers from 'various professional classes to the general public'. What had once been an audience has become 'a mass of protagonists'. 'We now have,' he writes, 'communication tools that are flexible enough to match our social capabilities'.

The cost of participation has also been lowered. Where once the decision to engage in political action required attending demos, finding petitions to sign and sitting through speeches, now anybody could sign up with a click of a mouse. Shirky also writes of the time freed for action and participation as people turned from passive TV watching to active engagement online. The engaged crowd was leading humanity to the sunlit uplands.

But by 2006, for Jaron Lanier, the crowd had become 'the hive mind', not a thoughtful mass of independent individuals but a blind collective driven by a desire to extirpate the human and hand all power to the internet. Lanier identified this process in the increasing number of 'meta' sites – Google, Wikipedia, news and blog aggregators Digg and Reddit, which aggregate from other aggregators, and, most notably, popurls.com, the supreme meta-site. 'We now are reading,' writes Lanier with wry dismay, 'what a collectivity algorithm derives from what other collectivity algorithms derived from what collectives chose from what a population of mostly amateur writers wrote anonymously.'

Then, in a passage that exactly echoes the anxieties of C. S. Lewis about the arrival of the men without chests, quasi robotic figures incapable of accepting any values or judgements beyond the most immediate impulses and demands, Lanier comes up with his own, internet-age version of the abolition of man. He links the apparent enthusiasm of the internet salesmen to eliminate human personality from the net to the race to produce artificial intelligence.

'In each case, there's a presumption that something like a distinct kin to individual human intelligence is either about to appear any minute, or has already appeared. The problem with that presumption is that people are all too willing to lower standards in order to make the purported newcomer appear smart. Just as people are willing to bend over backwards and make themselves stupid in order to make an AI interface appear smart ... so are they willing to become uncritical and dim in order to make meta-aggregator sites appear to be coherent.'

Our machines have become so seductive and effective that we want them to be our equal. To achieve this we have aspired to become less than we are, more simple, less complex. Lanier says we have inverted the logic of the Turing Test. We might think a machine is intelligent, not because it actually is, but because we have made ourselves more stupid to make it look smarter. After all, 'People degrade themselves in order to make machines seem smart all the time.'

No remark could be more carefully calibrated to offend against the assumptions of the Silicon Valley elite. In their eyes, our machines are making us smarter; they are fabulously effective prosthetics, extending the powers of the human mind. But, to Lanier, this is an illusion sustained by the fact that people are willing machines to be more than they are with such fervour that they are prepared to make themselves less.

But it would be wrong to identify this tendency as solely a product of the internet and Silicon Valley for those forces are, in fact, amplifications of the forces that already existed, notably in the increasing invasiveness and interactivity of the medium of television.

Simon Cowell was a successful music industry executive who became, in 2001, a judge in the first series of *Pop Idol* in the UK and, in 2002, in the first series of *American Idol*. Advised, apparently, by the publicist Max Clifford, from the first Cowell adopted the role of 'Mr Nasty', savagely criticising aspirant pop stars in a way that both outraged and delighted the audience. Cowell's appearance – very flat-sided haircut, dark jacket and white shirt with two or three buttons undone – added to the impression of a sharp, savvy businessman who knew what he wanted. The slightly strange hair apart, the jacket and the décolleté have become the standard uniform of young men wishing to look like hard-headed go-getters.

Cowell took few risks with the music. His favoured acts were all strictly middle of the road and designed to appeal to all ages, if not to all tastes. It worked. With his ensuing show *The X Factor*, now (autumn 2011) being launched in America, Cowell established himself as the most powerful man in the British pop industry.

These shows are, in fact, a twenty-first-century subset of the twentieth-century genre known as 'reality TV'. From its beginning as a mass medium in the 1940s, television has used the reactions of ordinary people as a source of documentary information or entertainment. So in *Candid Camera*, which began in the US in 1948, people were confronted with a strange situation – a lift that moves sideways, or a trick desk in which every time a drawer was closed another opened – and their reactions were secretly filmed.

The goal of Allen Funt, the show's deviser, was to trick people into acting naturally while performing for the microphone or the camera. This is, in essence, the goal of all reality TV. Subterfuge was necessary because of the extreme artificiality of the technology required. People froze in front of a microphone or camera as much as, or more than, they would freeze in front of a live audience. Funt's trick was to make them unaware of the machines. Later, 'fly on the wall' documentary film makers dealt with the artificiality of the situation, not by concealing the machinery but by habituating people to the cameras to the point where they started to ignore or forget them.

Later still, both the subterfuge and the habituation became unnecessary as people began to welcome the machinery as a normal part of their lives. Early TV clips of people waylaid in the streets by a crew and an interviewer show them tongue-tied and shy. Now every news show can find confident performers on any street to describe the scene of a disaster or to express their opinions. Being yourself on television has moved from an ordeal to an aspiration.

But, however normalised, reality TV remains a highly artificial form. There is nothing natural or real about appearing in front of a machine that will transmit your image to millions of people on camera. If people are habituated to the camera, then is that not simply a further form of artificiality? The parallel with Jaron

Lanier's anxieties about the way the internet has developed is clear – people adapt themselves all too willingly to the machines.

This adaptation was what made the new wave of reality TV show possible. This new phase began with *Big Brother*, which first appeared on Dutch television in 1999. It had been inspired by the 1991 Biosphere experiment in which eight people attempted to live for two years in an enclosed and sealed environment in the Arizona desert. This was a scientific project that ran into a series of technical and human problems. The failure did not discourage the TV producers. *Big Brother* simply took the idea of a group of people sealed off from the world. They were known as 'housemates' and were enclosed in a specially designed house. Their interactions as well as their responses to specific tasks were filmed. In 2002 the show was launched in Britain; it subsequently spread across the world.

BB's Dutch originators had intended to produce a series of discrete, more or less traditional TV shows which would, in effect, be collages of highlights. But it then became clear that there was an appetite for just watching the passage of time in the house. The psychological dramas of the enclosure combined with the audience's increasing familiarity with the characters had made banality engrossing. People started to become famous for the quality of their inactivity.

This was, of course, a version of famous for being famous, but it was a version with nuances; notably, there was the strange twist that enabled people to become more famous for being famous. By the autonomous logic of the combination of celebrity system and reality TV, people who have previously become famous for achieving something can become card-carrying celebrities simply by being made to do something unconnected to their expertise. So, in the show *I'm a Celebrity ... Get Me Out of Here!*, actors, sports stars and comedians are sent to survive and endure ordeals in The Jungle. The distinguished Conservative politician Ann Widdecombe appeared in the talent show *Strictly Come Dancing*;

the radical left-wing politician George Galloway, meanwhile, destroyed much of his credibility by appearing in the fourth series of *Big Brother* in 2006. The celebrity endowed by these appearances seems to cancel out any previous fame.

In Britain the most spectacular example of reality TV's form of famous for being famous – with absolutely no preceding attainment – was Jade Goody, an aggressively working-class girl who first made her name in the third series of *Big Brother*. Through a combination of startling gaps in her knowledge – she thought Rio de Janeiro was a person and wasn't sure what currency was used in Liverpool – frequently foul-mouthed and very politically incorrect remarks, the first sexual contact in the show, highly distinctive looks and a wild and often well-meaning enthusiasm about her growing fame, Goody became, simultaneously, a national heroine and villain. The tabloids called her first a pig and then a saint. She became an emanation from the masses, the ordinary people who, like her, spoke their minds and had no special gifts.

Goody's death from cervical cancer in 2009 seemed as bizarrely dramatic and brutally well timed as the death of Princess Diana twelve years earlier. Though both cancer and car crashes are primarily contingent events, both Goody and Diana were seen in death as defiantly heroic victims of their own celebrity.

I was at the funerals of both Goody and Diana and the mood among the crowd was identical. For all their faults, these women had defied their persecutors – the royal family, the hostile media, disdainful liberals – and attained in death a popular sanctity defined in opposition to that of the elites.

Goody was moulded by fame into something machine-readable – first as foul-mouthed renegade, then as brave, proletarian defier of authority. Throughout, it was clear she had no real idea she was playing either of these roles; rather she seemed to subscribe to her own sense of authenticity, an authenticity that was, in fact, defined by the roles thrust upon her. She was a victim all right, but not just of cancer or political correctness.

Simon Cowell now has his own rather more coherent Jade Goody in the form of Cheryl Cole, the celebrity who, as I mentioned in the previous chapter, has shown some awareness of the depersonalising machinery of contemporary fame. That did not prevent her being tossed about by that machinery when it was announced that she was going to be a judge on the American version of *The X Factor*. She arrived in America, comically badly dressed, only to be sacked, then reinstated, then sacked again. She had been set up to be a puppet clown born of the people but her strings were now being pulled, ultimately, by the machines that made reality TV possible and, in fact, necessary.

BB was connected to previous shows by these machines. Automated telephone systems, mobile phone texting, websites and the interactive button on digital TVs all made it possible for millions of people to vote. Since the phone calls were at premium rates, telephone voting also produced a new revenue stream to supplement advertising. These shows would not be possible but for the exponential increase in computer processing power and speed.

As marketing, this machine core was similar to interactive voice response systems in telephony. It transferred cost and inefficiency to the customer. The pop music business had previously relied on the ability of A & R (artists and repertoire) staff to seek out talent. This was an expensive, labour-intensive and uncertain process. The social media website Myspace, launched in 2003 (as MySpace), as well as the talent shows, made it clear that computer technology had rendered much of this process redundant. On Myspace, pop aspirants could post videos and sound files of their performances and the popularity of their music could be established by the simple metric of the number of hits on the website. Talent shows provide the industry with the same low-risk strategy of giving the public the ability to decide, in advance of the cost of recording and marketing, the likelihood – often, in fact, near certainty – of the financial viability of any given act.

That and their enormous success with viewers has led to a vast proliferation of TV talent shows across the world and to the global homogenisation of the style of pop music. Inevitably, as Simon Cowell had shrewdly understood, given the need to appeal to the demographics of television and to exploit the huge sample size provided by the voting system, the acts that succeed tend to be bland and impersonal, essentially averaged-out versions of the most instantly accessible forms of pop. The *Britain's Got Talent* contestant Susan Boyle may be seen as an exception because of the oddity, in this context, of her looks, character and background. In person, she is not bland, but her grotesquely over-cooked renditions of power ballads certainly are. Boyle is as aggressively middle of the road as it is possible to be.

It was this annexation of pop by a machine-driven system that led directly to the success of a bizarre anti-Cowell campaign. At Christmas 2009 Joe McElderry's 'The Climb' was number two in the UK Singles Chart. Number one was Rage Against the Machine's 'Killing In The Name'. 'The Climb' is an inspirational song – 'I got to be strong / Just keep pushing on' – 'Killing In The Name' is quite different – 'Uggh! / Yeah! Come on! Uggh! / Fuck you, I won't do what you tell me!'

Under normal circumstances McElderry would have been number one. He had won the sixth series of *The X Facto*r a week earlier and the popularity of the show – it had averaged a record thirteen million viewers in that year – would, without concerted opposition, have ensured more than enough sales to top the charts. It would have been inconceivable that he should be beaten by a rap metal band with a single which, in 1992, had only been twenty-fifth in the British chart.

But McElderry was singled out to be the victim of an uprising against Cowell's pop dominance. The campaign used the social network site Facebook and was orchestrated by Tracy and Jon Morter who said they were defying the Cowell 'music machine'. The Facebook group attracted 550,000 members. Cowell was angered. 'If there's a campaign,' he said, 'and I think the campaign's aimed directly at me, it's stupid. Me having a No. 1 record at Christmas is not going to change my life particularly. I think it's quite a cynical campaign geared at me . . .'

Later, however, he said he had been so impressed by the Morters' campaign that he had offered them a job. 'I now realise,' he added, 'I've taken too much for granted. I have got to hold my hands up. I accept there are people that don't like *The X Factor*.'

Even if they dislike *The X Factor*, few can resist the world of instant fame for ordinary people offered by Cowell's shows. In terms of the TV business, the great virtue of these shows is that they are a way in which this now ancient medium can compete with the internet. They are live, the dramas happen in real time. Recording them to cut out the ads destroys the whole point of the exercise; this is genuine event television.

But, for the wannabe famous, the TV gateway to celebrity remains narrow; it opens only to a few approved types. The internet is similar in its ability to generate machine-made fame, but it is wider and more accommodating. It is also infinitely more flexible when it comes to identity.

On 5 July 1993 the *New Yorker* published a cartoon by Peter Steiner showing a black dog sitting at a computer screen talking to a white dog with black spots sitting on the floor.

'On the Internet,' says the black dog, 'nobody knows you're a dog.'

It was, in 1993, amazingly prescient. Anonymity, the freedom to be who you pleased, was indeed to become one of the defining qualities of the new technology. Mario Armando Lavandeira Jr, for example, chooses to be Perez Hilton, celebrity blogger. 'Before me,' he tells me from his car on the way to the airport, 'most blogs were very first person, like online diaries – "I woke up today, I had tea with my neighbour, I went for a walk in the park" – blah, blah, blah, BORING! I didn't want to talk about myself. I wanted to talk about celebrities and entertainment because they're craaaaazzzzeee and they're so much fun.'

After graduating from New York University and trying acting, writing and PR, Hilton started blogging for which, it turned out, he had a real talent. His site was originally called Page-SixSixSix.com, a reference to the *New York Post*'s page six gossip column but with added satanic overtones in the form of 666, the Number of the Beast in the Book of Revelation. The *Post*, however, sued and the site became perezhilton.com. He also quickly landed himself in trouble with the actor Colin Farrell after posting a link to a notorious sex tape and with Universal Studios for posting a topless picture of Jennifer Aniston.

While still called PageSixSixSix, his site was named the most hated in Hollywood. This was good news for him as it meant he was read by the A list. It also meant he soon had 4.5 million visitors a day, a figure that rose to 6.1 million in the week when Anna Nicole Smith died and Britney Spears yo-yoed in and out of rehab.

Blogging made Hilton famous – though his gossip site has now been superseded by TMZ and RadarOnline – and blogging is the heart of the matter for both the most enthusiastic fans of the internet and of its fiercest critics. Blogging in the form of online diaries began in 1994. The very old media term 'diaries' was replaced in 1997 when the term 'weblog' was created and almost immediately shortened to 'blog'. But blogging did not attract mainstream interest until the first internet wave had crashed and was succeeded by Web 2.0 with greater interactivity and increasing numbers of broadband connections. The first reference I can find in the *New York Times* to the concept is a business story in April 2001. Blog at this point is still encased by inverted commas. By November 2002, however, the word had escaped the quote marks and the *NYT* was ready to offer 'Options for Beginning Bloggers'.

There are now so many blogs comprising the 'blogosphere' that counting them has become as difficult and redundant as counting trees or houses. In 2010 *The Blog Herald* said there were 147 million identified blogs, but the actual figure – if there is one – is likely to be much higher than this. The blogosphere is a true global crowd.

Blogs have two key attributes: anybody can easily start one and all posts are immediately published and made available to everybody in the world with online access. (This is not quite true in countries which subject the internet to political control, but it is both generally and in principle true.) So there are neither technological nor financial barriers to entry, nor are there distribution barriers to publication. In other words, if you want people to sample your talent, you do not need Simon Cowell or a publisher. You simply go online and set up a blog.

It is this circumvention of the gatekeepers of what bloggers call the MSM – mainstream media – that so excites the internet thinkers. 'The transfer of these capabilities from various professional classes,' argues Clay Shirky, 'is epochal, built on what the publisher Tim O'Reilly calls an "architecture of participation".'

Blogs are seen as a revolutionary tool that, potentially, draws the wisdom of the crowd into the fields of reportage, comment and political debate that had previously been jealously guarded by the professional media classes and rich corporate proprietors. Blogs are not just a new medium; they are a fundamentally different form of communication – freer and more democratic – that gives everybody everywhere an equal voice. But there is a third blog attribute that has proved central to the development of the blog form, the attribute captured by Peter Steiner's dog cartoon.

The internet allows you to conceal your identity. This is a very radical development. In ordinary discourse, people know who is speaking to them. This puts pressure on the speaker to take responsibility for what he says and to conform to the rules of the game by showing respect and a degree of deference. As an identifiable person, your status is at stake in all social interactions.

But internet anonymity removes all such restrictions.

Blogs have generated a new vocabulary for the strategy of anonymity. A 'troll' is somebody who posts deliberately provocative comments with the intention of generating anger or disruption of a discussion. 'Astroturfing' is when an organisation conducts an internet campaign – for advertising, political or public relations purposes – by concealing its identity to give the impression this is a spontaneous, grassroots behaviour. A 'sockpuppet' is an online identity designed to deceive, a mask to maintain anonymity.

The effect of anonymity is enhanced by the ease of the publishing mechanism – any fleeting, angry thought can be published in seconds – and by the scale of the internet itself. People are exposed to a larger range of opinion than ever before. This becomes hard to ignore.

Another cartoon – I traced it to www.xkcd.com – shows a stick man with a featureless head sitting at this computer.

'Are you coming to bed?' calls his wife.

'I can't, this is important.'

'What?'

'Someone is <u>wrong</u> on the internet.'

All of which creates a breeding ground for a form of behaviour which has been classified as a disease – mass psychogenic illness (MPI). Blogs infected by MPI are swept by strange and often fantastically cruel frenzies. Choi Jin-sil was found hanged in her shower by a rope made out of bandages on 2 October 2008. She was the most famous actress in South Korea and she had been hounded to death by rumours spread on the internet. Anonymous blog commenters had accused her of driving the actor Ahn Jae-hwan to suicide by demanding payment of debts. The wisdom of the crowd had become the savagery of the mob.

This points to a further way in which external regulation seems to be necessary if the crowd is, indeed, to be wise. First, as Surowiecki acknowledged and Wikipedia demonstrates, there needs to be regulation to prevent the crowd falling into factions. Secondly, there has to be some way of inoculating the crowd against MPI or mob behaviour. These requirements seem to suggest the wisdom of the crowd circles back to politics as usual in which an elite regulates the behaviour of the masses.

Yet, for the cyberprophets, politics as usual is now impossible. The internet, with its network effects and ease of access, its access to the wisdom of crowds and its circumvention of the mainstream media, represents a fundamental change in human existence. They free us to be able to live up to our highest destiny

'We now have communications tools,' writes Clay Shirky, 'flexible enough to match our social capabilities, and we are witnessing the rise of new ways of coordinating action that takes advantage of that change.'

'We are living,' he writes elsewhere, 'in the middle of the largest increase in expressive capability in the history of the human race.'

Across the Middle East in early 2011, Shirky's view seemed to be vindicated. After Mohamed Bouazizi, a Tunisian street vendor, was harassed and humiliated by the police, he set fire to himself in protest and, eighteen days later, he died. Thanks to the internet, Bouazizi's self-immolation set the whole region on fire. The Tunisian and Egyptian governments both fell, Bahraini leaders clung on thanks to massive concessions to the protesters and Libya and Syria became the latest territory to be subjected to the human penchant for massacres.

Young, technologically adept people were on the front line. Thanks to satellite TV, they saw different ways of life in other countries, they imagined change in their own and, using the internet and mobile phones, they organised. Facebook and Twitter were the tools of revolution – grateful Egyptians started to name their children Facebook. A Google executive, Wael Ghonim, emerged from eleven days of police detention to become both a hero and a leader of the revolt. In Libya, an internet shutdown was subverted by protesters who crossed the border into Egypt bearing flash drives from which they uploaded videos of state brutality. This, surely, was dramatic evidence that the crowd, democratic and wise, had been empowered.

Evgeny Morozov, a Belarus-born political scholar, is doubtful. Many attempts to bring down autocratic regimes using new technology have failed and, in time, tyrants learn how to use the internet as well or better than their opponents. The further twist, for Morozov, is that revolutionaries had better make sure they win. Internet and mobile communications are written in ink, not pencil, and the identities of their opponents will easily be traced by the regime.

Political change will happen thanks to technology, but it is too early to tell whether the change will be for the better. It is not too early to judge social and cultural change. Here, Jaron Lanier has concluded, change has definitely been for the worse. He speaks not merely in his role as digital apostate, but also in his role as a musician. In expecting something more, something finer, of the internet, he tends to focus on music as the primary example. Here the mania for everything to be free and instantly available has cut away the economic basis of creativity.

'In my view music is the canary down the coal mine for humanity. If the musical middle class is killed, then eventually the whole of the middle class will be killed.'

The musical middle class consists of the engineers, studio musicians and talent scouts who are being priced out of the industry by not just free and paid-for downloads but also by the abandonment of expertise and judgement in favour of the wisdom of the crowd. Lanier once expected the internet to create a new world in which tens of thousands of musicians would connect with millions via the web and, most importantly, make a good living out of it. Instead, the industry is collapsing on itself.

His anxious assault on the idea of the benign crowd is authoritative and ominous. If he is right, the internet as it is developing is potentially tyrannical and oligarchic rather than democratic. It is also destructive of the idea of art in popular culture in that it hands power to the dynamics of the mass vote, eliminating opinion and expertise. If he is right, then reality TV foreshadows a future in which creativity is controlled by the machine and the wisdom of the crowd kills art.

IT'S ALL IN THE GAME

'Computers,' says Eric Schmidt, 'make us better humans.'

Schmidt should know. He is a prince of Silicon Valley, a former director of Apple and chief executive officer of Google, a post from which he stepped down in 2011, though he retains the title of executive chairman. The enhancement of humanity is, among the princes, the unique selling property of their products.

'Take,' he continues, 'location-aware apps – for example, when you're walking on a street and your phone tells you that you need something from a store that you're walking by. That's the future. Imagine a near future where you never forget anything because computers, with your permission, remember everything.'

Later in the same speech he says, 'If you have a child . . . they'll have two states, asleep or online.'

Schmidt's future is Eli Pariser's filter bubble, a world of perpetual commercial surveillance and lives constantly nudged by devices that want to make us spend money. Every child, in this future, is, when conscious, online. As a result, Schmidt claims, they will be better humans.

Sherry Turkle, professor of the Social Studies of Science and Technology at MIT, may not agree that this makes us better humans. She has spoken to hundreds of children about their experiences of permanent connection. Sixteen-year-old Sanjay turned off his mobile at the beginning of his hour-long interview with Turkle. At the end he turned it on again and found himself confronted with a hundred text messages. His girlfriend was having 'a meltdown' and his friends were organising a concert so he was under serious pressure to reply. 'I can't imagine doing this when I get older,' he said as he left. 'How long do I have to continue doing this?'

Sanjay's phone seemed to be torturing him, loading him with obligations, absorbing him with connections he could not ignore.

Always to be online is always to be immersed in the demands of the online world. This is, to Turkle, troubling, but, to the Silicon Valley princes, 'immersive' is a thrilling, positive word. Software is at its best when it is immersive, when it seduces us into a condition of total and exclusive focus. It is a word marketing has gleefully embraced and there is now an immersiveness arms race with the word 'extreme' increasingly attached to signify the ultimate heights of cyberinvolvement.

Parents may think this is a good thing when software at school immerses their children in learning. After all, to be absorbed – or immersed – in a good book or a school project is taken as a sign that a child is doing well. But can it possibly be a good thing *all the time*? The very idea of a school education is based on the idea of immersion in many things. Reading Tolstoy alternates with maths and sports. The point of a liberal education is that the child returns intact from each immersion

But does he return intact when he never meets most of the people he calls his friends on Facebook or when he spends hours of his time playing games with children also spending hours in other rooms? How can any parent draw the line when each computerised sector – learning, socialising, gaming – seems to flow into the next? Who are they to say the skills acquired on *World of Warcraft* or Facebook will not prepare them for the future?

These are wholly new questions forced on parents by forms of technology that demand immersion in the virtual. Of course, as Turkle points out, the parents are unlikely to be wholly innocent, being themselves immersed in their BlackBerries or even their games – the average age of game players has been steadily rising – but, if they are parents at all in any meaningful sense, then the withdrawal of their children from the real world must at least strike them as odd. Somehow, they must find a way to calibrate levels of immersion to discover the point at which it becomes pathological.

From the appearance of the tennis game *Pong* in pubs and clubs from 1972 to the massively multiplayer online games (MMOG) of today, gaming is the thread that runs most clearly through the new age of total digital immersion. When the first personal computers appeared, nobody, not even the best and brightest, had any clear idea of what they would be used for. But they did at least know they would be used for games.

Nintendo is the company that has proved most successful at riding the successive waves of gaming fashion. I visited the company in Kyoto. Sitting in the lobby of the head office, I was first struck by the neutrality of the space. It was absolute, considered; there was no way of knowing what this company does. This is because, I was told, this is a place where business is done. It is not for customers so there is no need to display products. This is Japan and that division of function seems natural. In America or Britain such a lobby would probably be crowded with displays and invitations to know, buy from and participate in the company. In fact, as I stood in the lobby, an American visitor appeared. Loud-voiced and casually dressed, he stood out, a clear intrusion into all this exquisitely mannered austerity. To him, being in business meant self-assertion; to the men in dark suits it meant a purposeful self-immolation.

I was there to see Shigeru Miyamoto, who, at fifty-seven, was the greatest video games designer in the world. He was the man who gave us *Donkey Kong*, *Mario Bros*, *The Legend of Zelda*, *Wii Fit* and countless others. If influence can be measured in terms of sheer eyeball time, then he may well be the most influential man of the last twenty years.

His appearance was, in that extravagantly neutral space, shocking. Unlike all the men in that anonymous lobby, he was not dressed in neat black suit, white shirt and tie; he wore a black jacket and trousers which did not seem to be a suit, trainer-like black shoes and a t-shirt with a screen shot of an early *Super Mario* game. His hair was long. It used to be an early Beatles mop top, but now the fringe was swept back – more late Beatles. He looked like the wealthy drummer in a successful middle-of-the-road rock band with its roots in the Seventies.

Miyamoto and *Mario* saved this company. Nintendo was founded in 1889 as a maker of playing cards, but, by the 1950s, it had become clear that this was a very limited business. Nintendo diversified into love hotels, taxis and instant rice, all of which failed. In the 1970s it moved into electronics with a laser shooting system and home game consoles. Then, in what must be one of the most inspired hirings in corporate history, the company recruited Miyamoto. With a stock market valuation of $85 billion Nintendo is now the third most highly valued company in Japan.

At the heart of his success is the fact that he is an artist and industrial designer, not a programmer. 'The great majority of video games,' he said, 'are created by engineers. People like me who can draw some pictures were rather rare and I appear to be more important because of my rarity. I think I was simply lucky . . .'

Miyamoto's game style is a very Japanese mix of whimsy, fantasy and cuteness born of his eclectic immersion in *manga* comics, Disney and, from the UK, *Monty Python* and *Thunderbirds*. His games – all the iterations of *Mario*, *The Legend of Zelda*, *Wii Fit* and many others – are notable for their sheer friendliness. Much of the rest of the industry is dominated by war, blood and violence in games like *Grand Theft Auto*, *Resident Evil* and in MMOGs like *World of Warcraft*, *Blood Wars* and *Urban Dead*.

Consistently wary of generalisations and criticism of competitors, Miyamoto was reluctant to draw a clear line between these and his own games. His may be cute and very child-friendly, but they tell stories of conflict and they are, in their own way, very violent. But the violence is, somehow, filtered by the geniality of the visual style and the nobility of the tales.

'The most important thing,' he said, 'is substance rather than surface. I am not saying my games are totally harmless. I am always trying to speak to people about their fighting spirit. They don't want to be defeated, they want to win. For example, *Super Mario* is very violent . . . what matters is the manners, what kind of expression and description you use . . . I am always conscious of stimulating human beings one way or another. So, though my games can be played by children, they are not totally harmless.'

Stimulation is a warmer word than immersion. Immersion is inward looking; you are lost in the screen world. Stimulation implies that something about the experience can be transferred back into the real world.

His games divided into two phases. Phase one consisted of more or less conventional games of conflict, chases and races. Phase two, however, took the rapidly advancing technology in a wholly new and much more personal direction.

'I thought I would go back to the origin of myself and what I really wanted to do. I came to understand that I wanted to make an interesting toy that would be able to surprise people in a meaningful way.

'When I think about Nintendo it was just an entertainment company. And when I looked around the industry there were too many companies trying to compete against each other and outdo each other in the same categories.

'At the same time I had been growing up and now I had a family and I came to understand with my wife and my children that there is the living room in the centre of the household. Then I understood that we come together there and that digitally I could change the fun associated with that.'

Children change things. They made Miyamoto want to use his gifts in favour of real-world communality.

The Nintendo DS, launched in 2004, is a small, handheld,

double-screened console. It is wireless so players can interact. Over 132 million have been sold. The marketing is dominated by a community sense – people joining in – and by self-help – brain-training games made by the neuroscientist Ryuta Kawashima have been lapped up by the babyboomer generation who, in late middle age, keep losing their glasses and forgetting names. Now there is a 3D DS that does not need those glasses you have to put on in cinemas – a breakthrough that promised to galvanise the industry. The technology of 3D is definitely about increased immersion.

The demographics and dynamics of gaming have been changed for good by the Wii. The player's bodily movements – driving a car, hitting a ball – control the game. You can play almost any sport in your own living room and, another boomer dream, you can do your own workout. Almost seventy-four million Wii consoles have been sold and 573 million items of software.

Games, however warm, cuddly and Japanese, are another force driving us into the embrace of the machine and into becoming machines ourselves. Nintendo's extraordinarily intimate interface between human and machine was a sign of the future and it has been pushing competitors to ever greater heights of player immersion. Microsoft's Kinect, for example, requires no manual game controller. It scans the player's body and then produces a replica within the game on the screen. Real-world movements become virtually effective. This immersion is bodily as well as mental and represents a step towards a more total interaction between man and machine and a blurring of the line between the real and the virtual.

There is a similar blurring of the line between the real and the virtual in the larger scale sense that the virtual is taking on the political and economic dynamic of the real. *World of Warcraft* (*WOW*) – the first version appeared in 1994 – is the biggest of the MMOGs with more than twelve million subscribers. These players spend hours building communities, fighting wars and conducting byzantine diplomacy.

All of this is driven by a meta-economy. Money – known in the game as Gold – is required. This can be earned by spending hours and days doing boring tasks. But, thanks to Gold Farming, you don't have to. Chinese teams have formed to put in the hours – twelve hours a day, seven days a week in some cases – to earn Gold which they then sell to rich Western players for real-world money. (That is the explanation for all those junk emails offering WOW Gold.) A New York writer named Julian Dibbell spent a year making a living in this market, a project he documented in his book *Play Money: or, How I Quit My Day Job and Made Millions Trading Virtual Loot*. He pointed out, wryly, that the products of farmed Gold were much more robust than real-world financial derivatives.

Gold becoming real mirrors the Wii and Kinect feat of transforming real into virtual movements. Both are evidence of an increasingly intimate connection between the machine and the human, which, through gaming, is, in the minds of gaming prophets, about to change the world into an entirely gamed environment.

Pong, the video tennis game that effectively launched computer games on the world, handed control of the screen over to the viewers. Previously what happened on screens – cinema or TV – was something to be passively absorbed. This was at least four years before the Apple 1, the machine that signalled the start of the personal computer age and accustomed the world to the idea of the interactive screen.

For the next thirty years games were to drive the development of personal computers. They needed ever more processing power, memory and graphics, far more than basic functions like word processing, spread sheets and email. Manufacturers were under constant pressure to meet the demands of game designers.

Then, in the mid-1990s, a new genre emerged – artificial life (AL). In 1996 *Creatures* was launched. It had been created over a period of five years by Steve Grand, a self-taught English computer scientist. Grand's creatures were called Norns. They evolved, they had some kind of mind and they seemed to be able to do what they wanted.

'A game it may have been,' Grand later wrote, 'but if you'll forgive the staggering lack of modesty this implies, *Creatures* was probably the closest thing there has been to a new life form on this planet in four billion years.'

Creatures led to a wave of AL games – *Evolution*, *Sims* and, latterly, *Spore*. Typically these were 'God games' in which the player presided, godlike, over developments in the virtual world. *Spore*, for example, is based on the theory of panspermia, which argues that life on earth was seeded from outer space. The game is a new form of MMOG. It is played alone, but created by its users. In *Spore* you can make a creature which will then be uploaded to appear in the games of others. *Spore* is a collective task, a product of the 'meta-brain'.

'Any human institutional system,' says *Spore* creator Will Wright, 'that draws on the intelligence of all its members is a meta-brain. Up to now, we have had high friction between the neurones of the meta-brain; technology is lowering that friction tremendously. Computers are allowing us to aggregate our intelligence in ways that were never possible before.

'If you look at *Spore*, people are making this stuff, and computers collect it, then decide who to send it to. The computer is the broker. What they are really exploring is the collective creativity of millions of people. They are aggregating human intelligence into a system that is more powerful than we thought artificial intelligence was going to be.'

Games, according to Wright, may achieve what AI researchers have failed to do – create an autonomous intelligence – through the wisdom and joint work of crowds rather than through the research of computer scientists. This is another version of the belief of Greg Riker at Microsoft. He thought a different species would emerge from this new technology, a species with brains expanded by their contact with machines. But he could have had no inkling that this new species might game its way into existence.

By the mid-1990s, people were beginning to feel queasy about the growth and effects of computer games. In his presidential address to the Royal Society in 1995, the mathematician Sir Michael Atiyah said: 'I find it an odd reflection on our society that some of the most sophisticated technology, resting on the contributions of our greatest intellects, finds its ultimate destiny in computer games.' Atiyah plainly found something trivial or unworthy about such an application of intellect. Games were not as important as ending poverty or curing cancer.

The greater queasiness came from the feeling that there was something actually wicked about these games. In 1999 Dylan Klebold and Eric Harris walked into Columbine High School, Colorado, and killed twelve students, one teacher and themselves. Klebold and Harris, it was said, derived their firing and targeting techniques directly from *Doom*. This game, which came out in 1993, is a 'first-person shooter' which means you, the player, shoot at your enemies. The connection was made between virtual and real violence.

Later came concerns about people's high level of absorption in video games. In 2010 South Korean police arrested a couple – a forty-one-year-old man and a twenty-five-year-old woman. Their three-month-old baby daughter had died of starvation. Her parents had been too busy bringing up a virtual daughter in a video game to bother feeding her. A twenty-two-year-old Korean man was charged with murdering his mother because she nagged him about his excessive game playing. In 2005, again in Korea, a man collapsed in an internet café and subsequently died after playing *StarCraft*, a sci-fi war game, for fifty hours.

All games exist in an unreal hinterland which can become real. Those people really died. It is this interaction between real and the unreal that has, throughout history, made us aware of the fundamental importance of playing games. Game-like calculation in war, for example, goes back to the ancient Greeks and, most spectacularly, to the sixteenth-century Spanish conquistador Hernán Cortés. Having landed in Mexico and found himself potentially outnumbered by Aztec forces, Cortés stripped and scuttled his ships. This served two purposes. It gave his soldiers no choice but to fight. This is important in game theory because there is always a danger in war that individual soldiers may reason that their presence will make no difference to the outcome and choose flight rather than fight. The gesture also suggested to the Aztecs that Cortés was confident he could not lose. In strict battlefield terms, Cortés's gesture made no sense, but in abstract, strategic terms – game terms – it was a brilliant move, like a winning sacrifice in chess.

The reasoning of Cortés, his soldiers and the Aztecs is like mathematics, an abstraction that, somehow, describes and can make forecasts about the real world. They are all agents, creatures with preferences – to win, to live, to flee, to profit – who both make calculations and who, crucially, are subject to the calculations of others. In the mid-twentieth century the full significance of this seemed to become apparent.

Mathematical giants – including John von Neumann and John Forbes Nash (whose story was told in the film *A Beautiful Mind* (2001)) – began to study games. What made the subject so intriguingly difficult was the idea of interactions between agents. If I kick a ball, the outcome is straightforward – it moves; the ball has no preferences. If I kick you, the outcome is subject to a huge proliferation of possibilities – will you kick back, fight, run away, sue me? – all dependent on the fact that you are also an agent with preferences.

The most celebrated demonstration of how this might be turned into mathematics was the game *Prisoner's Dilemma*, devised in the 1950s. Two men, suspected of a crime, are arrested. The police do not have enough evidence so they separately offer each man a deal. If one testifies against the other and the other says nothing, then the testifier goes free and the silent one gets a full ten-year sentence. If both remain silent, they both get a six-month term. If each betrays the other, both get a five-year sentence. So betrayal or silence are the two choices. There is no contact between the two suspects. What is the rational thing to do?

The strategies can be mapped mathematically easily enough. The rational choice turns out to be betrayal because this removes the risk of the full sentence and offers the chance of immediate freedom. This is, of course, to assume that the betrayer cares nothing for the betrayed. But, if the game is played repeatedly and each player is made aware of the other's previous behaviour, then a new psychological factor emerges. They may both decide to remain silent and take the six-month sentence.

The mathematics of game theory led to an explosion of game theoretical approaches to economics, diplomacy, sociology, engineering, computer science and philosophy. Game theory, it was claimed, was a 'universal language for the unification of the behavioral sciences'. In media, management and government circles today, people routinely talk about 'gaming' a problem in order to test out all possible outcomes. Clever operators are often said to be 'gaming' the system.

But it was in biology that game theory's true significance seemed to emerge. By the mid-1950s all the elements of what we now call Neo-Darwinism were in place. A hundred years earlier Charles Darwin had produced his theory of natural selection to explain the development of living species. It is all a game. Plants and animals compete for limited resources and are occasionally rewarded by a favourable genetic mutation which gives them a competitive edge. Their moves in this game are intricately interwoven with the moves of others. It was simple, understandable to the layman, and it seemed, to many, irrefutable. But it took a century to underpin it with hard science. The climax of this process was the deciphering of the molecule of DNA in 1953. Finally we had what Darwin did not have but which his theory desperately needed – the mechanism for transmission of information down the generations.

But a gaping hole remained – altruism. Why are people nice to each other? Why do animals sometimes act against their own best interests? If Darwinism is correct, then it would seem each creature's best strategy would be, in terms of *Prisoner's Dilemma*, to betray. But we don't and, in fact, animals don't.

Game theory suggests there may be a solution. Evolution not only determines the physical attributes of an organism, it also determines its behaviour. If an animal does well by a particular strategy for seeking food, it will live longer and reproduce more. In a competitive environment – and *all* environments are competitive – a key aspect of evolved behaviour is how it impacts others. As *Prisoner's Dilemma* shows, this may mean a constant strategy of betrayal or it could mean cooperation. In fact, after much computer churning, one very stable strategy was discovered called benign tit-for-tat.

In every encounter you should make the initial assumption that the other is a cooperator. If he turns out not to be, then you respond by non-cooperation. If he is, then you continue to cooperate. In the computer simulations this produced a robust and stable society. Altruism suddenly popped out of the programmes. It was not God-given, it was just there in the maths. Life was a game play.

This idea gained further credibility in 1970 when the British mathematician John Conway came up with *The Game of Life*. This was a computer game of almost bewildering simplicity. Technically it is known as a cellular automaton. Imagine a grid of squares (cells), some black, some white or some 'on' or 'off'. Now you apply a set of rules – for example, if a black cell is surrounded on three sides by whites, then it turns white. In *The Game of Life* the cells were either alive or dead and subjected to four simple rules, each of which had some evolutionary resonance. Fewer than two live neighbours will result in the death of a cell; more than three live neighbours and the cell dies; with two or three live neighbours it lives on; a dead cell with three live neighbours comes back to life.

The first move is to turn on the game. There are no other moves, it runs itself. A series of patterns emerged – they were called gliders, glider guns, beehives, pulsars, toads and boats. A procreative, self-contained world emerged as if by magic. This did, indeed, appear to be the game of life and it could be modelled on a computer.

If from the bottom up the world is a game, then it is also becoming one from the top down. Prophets of an entirely gamed human future are beginning to emerge from the highest ranks of the technocracy. There are two themes here. For some, the millions who are now acquiring game skills represent a gigantic new meta-brain whose intelligence can be applied to solving the large-scale problems of the real world – poverty, global warming, war. For others, the ordinary quandaries of day-to-day life can be dealt with by turning them into games.

The latter group can reasonably claim this is already happening. In Britain, for example, schoolchildren are encouraged to walk to school by the Walk Once a Week scheme. If children walk at least four times a month, they earn a badge designed to become collectable. The agent, the child, is given an incentive to overcome its reluctance to make the effort.

At earndit.com you can join various programmes also intended to persuade you to exercise more. Their analysis is based on the concept of 'hyperbolic discounting'. People tend to discount the value of any reward that lies far in the future. Exercise is the perfect example. The benefits may take months to be felt or seen so there is always an excuse not to exercise now. The incentive at earndit is monetary – you earn points per mile of walking, running, cycling or for every gym check-in. Cheating is prevented by the use of various devices like the Nike+iPod or Garmin GPS. The system works, in effect, like a store loyalty card.

At ultrinsic.com students can bet on their college grades. Based on how much they bet and on their previous performance, a cash reward will be calculated. At usvirginhealthmiles.com, employers are provided with an incentive system to get employees to exercise and thereby reduce healthcare costs.

The Tooth Tunes toothbrush, meanwhile, encourages 'better brushing habits' by playing your favourite song while you brush. The sound vibrations are transmitted through the bristles so that you hear the songs through the bones in your head. Increasing brushing pressure increases volume and you hear the full two minutes of music if you keep brushing. In Brazil some Omo detergent boxes contain a GPS device. This provides the company with details of their location. They are called on at home and given a prize.

'Advergaming' is the clumsy title given to these developments. Part of the background to this is the widespread feeling that conventional advertising is in its death throes. It is seen as an aspect of 'old media'. Advertising treats the consumer as a passive receptor. But, in the world of new media, the consumer expects to participate through loyalty cards, through various mechanisms to make them feel part of a club and, increasingly, through games – hence advergaming. Fun is the goal, say Gabe Zichermann and Joselin Linder in their book *Game-Based Marketing*.

We are approaching, says Jesse Schell, the Gamepocalypse. This seepage of game thinking into ever more areas of ordinary life is celebrated by Schell of Schell Games as a ludic version of Ray Kurzweil's Singularity. Schell makes use of the example of *Farmville*, a farm simulation game on Facebook. It is the site's most popular application with sixty-two million users.

'Look for nooks and crannies – find the spaces no one else has made a game for,' he says, 'a big part of the success of games like *Farmville* is that you can play them at work and the boss doesn't notice. How can I get games that happen when I'm cleaning my house? Is there a game that can help me fall asleep?'

In other words – games in all things at all times. This is now rapidly being formalised into a globally ambitious ideology of gaming.

On her blog, Jane McGonigal says she 'makes games that give a damn'.

'I study how games change lives. I spend a lot of time figuring out how the games we play today shape our real-world future. And so I'm trying to make sure that a game developer wins a Nobel Prize by 2032.'

In an extraordinary presentation at a TED – Technology, Entertainment and Design – conference in February 2010 Jane McGonigal established herself as the leading prophet of the idea of games always and everywhere and, specifically, of the belief that they can solve big real-world problems. She is a Ph.D. in performance studies, Director of Game Research and Development at the Institute for the Future, a Palo Alto think-tank which emerged from the RAND Corporation. She developed *Superstruct*, an MMOG based on the idea that the human species has only twenty-three years left, and *World Without Oil*, an alternative reality game (ARG) designed to come up with solutions for an oil-free future.

McGonigal's TED speech was all about her plan 'to make it as easy to save the world in real life as it is to save the world in online games'. She pointed out that three billion hours a week are currently spent playing online games, but this was 'not nearly enough'. We need to raise that figure to twenty-one billion hours to solve problems like poverty and hunger.

Gamers feel they are better at games than they are at reality – 'In a game world many of us become the best version of ourselves.' The trick, therefore, is to apply these better versions to the real world. 'It turns out,' McGonigal said, 'that by spending all this time playing games we are changing what we are capable of as human beings. We are evolving to be a more collaborative and hardy species.'

Again there is that Silicon Valley theme of human enhancement through gadgetry.

The average young person has spent 10,000 hours playing online games by the age of twenty-one and has thus become a master of a whole range of skills that could be applied to the real world. Such advanced gaming skills – involving, crucially, collective problem solving – give gamers four attributes: urgent optimism, social fabric virtuosity, blissful productivity and epic meaning. They are, for McGonigal, super-empowered people.

Given that, in the words of the economist Edward Castronova, 'we are witnessing what amounts to no less than a mass exodus to virtual worlds and online game environments', it seems reasonable to make these games useful in the real world. If we do, according to McGonigal, 'we can make any future we can imagine and we can play any games we want'.

There were signs of incredulity and some laughter in the course of the speech. McGonigal seemed to revel in this, apparently feeling validated by the obvious scepticism. Plainly the doubts of Sir Michael Atiyah about the expenditure of so much intellectual effort on games persist even in these exalted technocratic circles. But, equally plainly, this scepticism is being overwhelmed by the numbers – millions of players and billions of hours.

McGonigal has now established a 'secret HQ' – actually a website called gameful.org – for people interesting in making humans 'happier, smarter, stronger, healthier, more collaborative, more creative, better connected to our friends and family, and better at WHATEVER we love to do when we are not playing games'. Games, in this vision, are not an escape from the real; they are a way of returning the now super-empowered to the real.

I find all of this hard to understand. I loathe computer games and feel guilty and gloomy if I ever find myself, however slightly, addicted even to simple ones like *Tetris*. But, out there at the technocratic cutting edge, the game has emerged as a new paradigm, a new way of understanding and changing the world. From the mysterious ways of cellular automata and the proliferating complications of game theory to health incentive schemes and game marketing, the idea of the game has intruded into all aspects both of the human and the non-human world. Whether it is nothing more than just a new way of looking at things or the discovery of a fundamental truth about reality remains to be seen. What is clear, however, is that the games are further masks for the machines.

There is a third alternative – that gaming is a new way of looking at things that we can turn into a fundamental truth. If the world in our heads becomes a game, what need would we have of the world outside our heads? We can decide the game is truth.

Perhaps, the physicist Paul Davies muses in his book *The Eerie Silence: Are We Alone in the Universe?*, this has already happened. Davies imagines the possibility of technologically super-advanced aliens who have turned themselves into quantum computers and become extraterrestrial quantum computers (EQCs). Such beings would want to move to interstellar or even intergalactic space – it would be the coldest possible environment and, therefore, the most stable for quantum computers. What would they do when they got there? They would retreat into the game of virtuality. The truth of their world would exclude the physical cosmos. Is this where all this might lead?

Shigeru Miyamoto warned against all this in his diffident way. He said his games are not made to absorb people within the game but to create something that 'expands beyond the world of the game'. He accepted there have been terrible cases of game addiction, but, as millions play video games, extreme cases are inevitable. Also there are moral panics about all new technologies. Television was frequently blamed for destroying lives and cultures.

On the other hand, far greater claims are being made about video and computer games than were ever made about any previous technologies. They are demanding, intimate, addictive, unreal or, in McGonigal's terms, world transforming. But are they just another simple solution to a complex problem? Will they turn humans into prisoners of the virtual?

'How long,' asked Sherry Turkle's interviewee Sanjay when confronted with a hundred text messages, 'do I have to continue doing this?'

Money – the making of it – is often seen as a game. In the form of what used to be called 'high' finance, it is, like a game, an abstraction, a hypothetical version of reality. Of course, in finance the link with the real world is more direct. Projects great and small and people's livelihoods depend on the workings of the financial system. Nevertheless, financiers themselves are very easily seduced away from the real. In 2008 it became clear that they had abandoned all contact with reality and lost themselves in a game of machines and fantasy mathematics.

Paul Wilmott is a quant, meaning he applies mathematics to finance and investment. Nassim Nicholas Taleb, who wrote *The Black Swan*, is also a quant and says Wilmott is among the best in the business: 'Paul Wilmott is the smartest of the quants ... he may be the only smart quant.'

Wilmott may also be the most colourful quant. He is fifty years old, but looks much younger. He publishes *Wilmott*, probably the most expensive magazine in the world. He had forgotten how much it cost when we first met. I told him it was about £600 for six issues but I was wrong; it is, in fact, a fairly reasonable £395. He also makes cheese. The fridge in his flat contains industrial quantities and multiple varieties of home-made cheese.

He is a charismatic speaker, demanding participation from his audience in cleverly devised problem-solving tasks. He especially likes making his points through stories, jokes and imaginary conjuring tricks. This is in sharp contrast to the style of his friend

Taleb, who is a much more introspective speaker, expecting rather than demanding his audience's attention.

Wilmott and Taleb have jointly been running a series of seminars in London. They have done nineteen since 2003. These cost £1,999 per person for a two-day course and they are attended by risk managers, traders and quants from hedge funds and banks and, occasionally, some private investors. The seminars are entitled 'Robust Risk Management'. RRM is an idea that goes beyond ordinary quant thinking because it springs from the conviction that the mathematical models of how investments will behave may not be enough. 'It's always been about errors in the models,' Wilmott explains, 'looking for robustness rather than fancy maths ... especially how markets behave in crashes and extreme situations.'

As everybody learned during the banking crisis of 2008, mathematics, no matter how sophisticated, is often incapable of capturing the reality of financial markets. Maths is a ridiculously simple solution to an obviously complex problem. RRM is all about knowing this and learning to expect the unexpected. Markets are both abstract in the sense that they are made of numbers and human in that they are driven by sentiment and error. They are complex and, therefore, cannot fully be captured by the computerised models of the quants. Belief in the simplifications of these models lay behind the 2008 crash which, as a result, has come to represent one of the most vivid and damaging examples of the dangers of attempting to apply simple rules to a complex system and of submitting ourselves to the logic of the machine.

In contrast to the simplifiers, Taleb and Wilmott are artists of finance. Like all great artists, they attempt to encompass the unknowable and the uncertain in their work and to respect complexity. Few people are able fully to grasp the radical significance of what they are saying.

The seminars are difficult or impossible for the layman to

understand. The one I attended was in an anonymous office block on the edge of the financial district in Hatton Garden which has, since the Middle Ages, been the centre of London's jewellery trade. Taleb and Wilmott take turns to speak to an audience of young, earnest, ambitious but slightly sceptical high-flyers. They argue with both speakers and it is unclear whether they have been converted to the creed of RRM or are still clinging on to the more conventional – and risky – wisdom of the City. Perhaps it is a little of both.

Yet, even for the layman, the seminars possess a weird but accessible drama because of the clashing personalities of the two protagonists. Wilmott is a dramatist, constantly coming up with real-world stories to illustrate his points. He always tries to go for the concrete so, for example, he does not talk about holding securities, he talks about 'putting them in the bottom drawer of your desk'. Taleb has an introverted style of delivery and he speaks with an epic impatience with the follies of the financial world. For him, there is no need for the drawer in the desk; his numbers and charts speak for themselves.

This points to their one deep intellectual difference. Wilmott thinks the mathematical models used by the banks and funds are viable ways of handling investments in certain carefully defined situations. Taleb thinks they are simply wrong and their inventors should be stripped of their Nobel Prizes. But, on the general principle that the mathematics of investment do not constitute a hard science, they are both agreed. They are also both agreed on the incorrigible consistency of human folly.

Wilmott was born in Birkenhead, across the River Mersey from Liverpool, and he retains a slight Scouse accent. He was mathematically gifted but he disliked pure maths, which he seemed to see as somehow disordered. Pure maths is simply the pursuit of mathematics with no application in the real world. Wilmott can see no point to this, preferring applied maths, the use of numbers to control or change the real world. He is very proud that somebody once said of him that he was 'the only normal mathematician I have ever met'; it represented an endorsement of his own connection to the world outside numbers.

At St Catherine's College, Oxford, where he earned his degree and doctorate, he worked on practical subjects like fluid mechanics, submarine design, aircraft wing design and some 'weird stuff' like the mathematics of how to shave. I will return to shaving later.

In 1987 he came across the equations then used in finance and investment. To his amazement he recognised the style of the maths. 'I assumed that there wasn't any mathematics in finance but then I saw this famous formula, called the options pricing formula. The equations were very similar to the ones I had been working with since I was eighteen and now I was seeing them in finance. I started dabbling in finance and then, over the years, I dropped other forms of mathematics and finance took over.'

The options pricing formula is known as Black–Scholes, or Black–Scholes–Merton (BSM). It first appeared in a paper published in 1973 entitled 'The Pricing of Options and Corporate Liabilities'. It was written by two economists, Fischer Black and Myron Scholes. Another economist, Robert Merton, developed the idea and it became the Black–Scholes options pricing model. Merton and Scholes won the Nobel Prize in economics in 1997 for their work. Black had died in 1995 and was therefore ineligible.

Black–Scholes seemed to solve what had been the most difficult problem in finance. An option is a right to buy or sell an asset – a commodity or a share, for example – at a certain price at some point in the future. The price of an option is based on the current price but this is combined with a premium based on the time remaining until the expiration date at which point the option becomes worthless. Such deals are extremely complicated to model mathematically. BSM was the solution, but, as we learned in 2008, it was also an entirely new problem.

What Wilmott first noticed about the equations was not money but structure. Equations have differences in style and form that, for mathematicians, are as clear as the difference between the music of Mozart and Lady Gaga.

'When you see an equation,' explains Wilmott, 'you see the structure and you can say, "Oh that equation, it looks a bit like this one". The mathematics of finance or derivatives in particular, those equations are very similar to the equations of heat flow which go back two hundred years or diffusion which goes back to 1905 and Einstein again.'

Wilmott knew how to be a quant before he knew there was such a thing and he enjoyed himself experimenting with finance as just another field of mathematics. 'In those days you could do any kind of maths you wanted and people would respect you as long as two plus two equalled four and they felt you had something to contribute. And then, about fifteen years ago, quantitative finance changed completely.'

What changed was that financial mathematics was suddenly elevated to the role of hard science. The connection Wilmott had seen between his own applied equations and those of the quants was significant but dangerously misleading. He had always been aware that, though the equations of finance used the tools of science, it lacked the underlying principles of science. In finance, there are no basic, real-world laws to which the equations refer.

Physicists know that their speculations are anchored by solid, observable truths like the inverse square law of gravity or the law of the conservation of linear momentum. (There is a problem about what these laws actually are – embedded in reality or in our maths – but they form a reliable foundation for scientific inquiry.) Market traders have no such laws. Just as the price of an option – and all such 'derivatives' as they are known – is disconnected from the oil, wheat or shares in Microsoft on which it is based, so the mathematics of finance have no physics to anchor them to the real world. Certainly they work – some of the time – but they do not work because they represent laws of nature;

they just happen to work in certain circumstances at certain times. They are complicated instruments but they cannot deal with the true complexity of the real world.

This should be obvious. A market is not a planetary or a sub-atomic system. It is an arbitrarily defined space in which human beings interact. Any market may be subject to external or internal shocks which cannot be captured in equations based purely on the logic of the market. But this was not obvious to the new wave of market mathematicians trained and trading in the 1990s.

'Universities realised,' says Wilmott, 'there was money to be made in finance. They started teaching people master's degrees in financial engineering. And then suddenly instead of having all these possible worlds of mathematics, they decided this is how you do finance. I swear I could stand up and start talking about my kind of mathematics and these people, from this very narrow academic perspective, would think I was an idiot. Instead of being a subject where anybody could contribute, it became a subject where you've got to do it this way or else.

'The reason why I think this happened was because there were a lot of people in the universities who had no experience of the real world; they started putting on mathematical finance programmes without any real-world applications. So these poor twenty-two-year-olds came out of universities with their degree and they never had any experience of the real world. They see that money is being made so they sign up for these $80,000 courses.'

Those twenty-two-year-olds became the quants who played such a crucial role in the market boom that led to one of the worst crashes of all time in 2008. They are far from being the sole culprits but they did propagate one of the key illusions that led to the crash – that machines, fed with certain types of mathematics, could control the world. This is a gross and demonstrably false simplification.

They had been deluded into thinking that their equations had the same experimentally verifiable reality as Newton's or Einstein's. But the truth was that they were simply useful tools that only worked within very narrow circumstances and very limited time frames. This raises the question – can markets ever be captured by equations?

For the moment, we know that they cannot. In fact, this had been known for some time before 2008. The limitations of these tools had repeatedly been demonstrated, most dramatically in the case of Long-Term Capital Management (LTCM), a hedge fund founded in 1994 with, on its board of directors, Myron Scholes and Robert Merton, two of the gods of quantitative analysis. LTCM delivered 40 per cent returns for a while and then, in 1998, it crashed, losing $4.6 billion in a few months and threatening the stability of the entire US financial system. It had been destroyed by the Russian debt crisis, what Taleb would call a black swan, which did not – could not – appear in the Black–Scholes formula.

'Obviously,' commented Scholes after the crash, 'you prefer not to have lost for investors.'

Merton, in an interview in Wilmott's pricey magazine, defended BSM by saying it was like a four-wheel-drive car. In snow, the four-wheel-drive offers improved safety, but this lures drivers into taking greater risks so there are just as many crashes. BSM is four-wheel-drive for the markets, but, like any car, it will go out of control if pushed too hard.

In spite of LTCM and many other failures of quantitative analysis, the quants continued to believe in the scientific basis of their discipline. In fact, they still do so with quasi-religious fervour. In his 2010 book *The Quants: How a New Breed of Math Whizzes Conquered Wall Street and Nearly Destroyed It*, Scott Patterson reveals the quants' obsession with something they call The Truth. This would be the ultimate mathematical statement about markets and they thought they could find it in their equations applied with ever-increasing computing power. 'The bigger the machine,' Patterson writes, 'the more Truth they knew, and the more Truth they knew, the more they could bet.'

Quants, as he explained to me, are reluctant to acknowledge that The Truth – sometimes they call it Alpha – is their Holy Grail. 'This is something you're not going to find in any quant literature, but when you talk to them privately they'll use the words The Truth ... It is a cabalistic search for the nature of human beings and how they interact and the market's reflection of that. They believe in mathematics, they believe their formulas will be able to capture it and they'll make a load of money off it.'

This superstitious belief in The Truth is a very significant development in our machine history. Computer modelling is the first new way of doing science since Galileo. It adds a third method to experiment and observation. With computer models, we can create virtual worlds which, if sufficiently carefully constructed, can cast entirely new light on the real one. In other words we are not limited to what we can see or detect; we can observe and test *possibilities*.

But, precisely because of this power, computer models have to be handled with care. They are, inevitably, approximations simply because no human can imagine nor any computer contain all the variables in the real world. This means any model has to be rigorously restricted in its scope and its design has to be very precise if results are to be any use at all. No serious user of computer models would ever claim that they contained anything as final or as complete as The Truth.

But the quants seem to be mesmerised by this spectre. This produces a bizarre state of mind in which mere complication in mathematics convinces quants that they really can control risk. Wilmott points out that quants thought they understood CDOs – the collateralized debt obligations that played such a huge part in the crash – simply because of the complicated mathematics. When combined with bank managers who did not even make a pretence of understanding them, this produced the belief that the CDO market, and all derivative markets, could be expanded without affecting the underlying risk calculations. In 2008, at the height of the crisis, the total value of all derivatives was estimated to be between $600 and $700 trillion, or twelve times more than the whole world economy.

Wilmott has been made immune to the superstitions of the quants by his natural scepticism. Also he is very risk averse. He doesn't gamble but he sees that gambling with their own money would be a useful thought experiment for bankers. 'I don't gamble, I am very, very risk averse. I have always thought people who work in banks should gamble with their own money in casinos and get more knowledge about themselves. I ask audiences whether they would take more risks with other people's money than with their own and some say of course; some are more responsible and say they wouldn't. People ought to know more about their own responses and they would find out by gambling.'

The quants' superstition that infected the financial markets is the most vivid example of a superstition that infected the entire world in the post-war period – the fantasy that maths could be applied to the human realm and, with the ever-increasing power of computers, arrive at truths that were as hard and testable as those of physics.

In economics in particular, the market fundamentalism that sprang from the works of Milton Friedman was seen as the beginnings of a final statement about the nature of the human world. As such it covered not just investment but also morality. If this was a truth of the world, then it was morally incumbent on the individual to act in accordance with that truth. This was the assumption that lay behind the line 'Greed is good' uttered by Gordon Gekko (Michael Douglas) in the movie *Wall Street* (1987). The implication was that the moral injunction followed in a law-like manner from the underlying truth of the market. The individual's pursuit of wealth would inevitably enrich us all, therefore greed was not an excess, it was an obligation. The bottom line of the profit and loss account was also the top line of moral behaviour.

But some things – most things – humans do are nothing to do with the bottom line. Cheese, for example. Cheese has probably been produced since humans first became settled farmers 10,000 years ago. It can be seen as both a natural and a highly artificial food – natural because it simply happens when a protein in milk coagulates, artificial in that it requires tame, milk-delivering animals and a reasonably stable community to allow time for the cheese to develop. The passing of time – ageing – is important in most cheese production. Cheese is a complex product of the interaction of man and nature.

Paul Wilmott, as I said, makes cheese. 'The catalyst was a trip to a monastery in Switzerland (a skiing holiday during which I stayed in cafés at the tops of mountains and read) which made cheese. And I bought a child's home cheese-making kit. I now have all the proper gear, have made hard, soft, cows', goats', pressed, non-pressed . . . Currently I have a lethal-looking Stilton ripening for Xmas. My enthusiasm for learning new things goes back to childhood. And historically I've often turned a hobby into a business. However, I can't see this happening with the cheese making: margins far too small!'

The cheese hoard in Paul Wilmott's fridge would alarm Arthur de Vany. De Vany, Professor Emeritus of Economics at the University of California, Irvine, was seventy-one when I met him at his home in Utah in 2008 but, since he was wearing only a towel when he opened the door, I could see he had the body of a very fit man in his thirties.

I was visiting him because a few weeks before I had spent two days interviewing Nassim Nicholas Taleb in Newport Beach, California. Our conversation was supposed to be about finance, economics and human folly in general, but, in the event, much of it turned out to be about diet. Taleb had adopted a very low-carbohydrate regime based on the probable eating habits of humans before the advent of farming. Plainly such a diet could not include cheese. This was, he argued, the way we were meant to eat because it is how humans sustained themselves throughout most of our evolutionary history. Pre-agricultural humans did not eat cheese, nor did they eat cereals, potatoes, pasta and rice. Taleb had adopted the diet after reading and talking to de Vany.

I also adopted the diet – apart from the no-cheese part which I find intolerable – after meeting Taleb. The effect was rapid. In two weeks I had not only lost weight but I looked better. So much better, in fact, that I went to Utah to interview de Vany. He was ready with all the evidence for his diet from within his own body.

'I'm not on any medication. I had a fair number of colds before I began this programme but I've only had two episodes of food poisoning since then – both were in high-end restaurants – and that's it since 1984. My insulin is unmeasurably low; insulin is the ageing hormone, it tells you to go ahead, reproduce and die. My HDL [good cholesterol] is enormously high, my triglycerides are way down, my blood pressure is perfect, scans of my carotid artery show there are no lesions there, no build-up of plaque, so my brain is getting lots of nourishment. I was just motorcycle riding with a gang of guys in Elba, riding all over the hills on my motorcycles. I play softball – I was the only guy in the seniors to hit them over the fence. I can hit a golf ball 340 yards. I'm fast as heck, I'm very strong.'

Economics happens inside as well as outside the human body. His work is all about the dynamics of complex, adaptive systems; he is a complexity scientist which means he studies how systems work rather than breaking them down into their component parts. Central to this is the overthrow of old statistical models. Basically, we have all been taught, that events – human wealth, earthquakes, blockbuster movies – cluster round an average forming a graph in the shape of a bell curve. This is an illusion and the concept of the average leads to fatal errors.

'The average,' de Vany said, 'is always misleading and may not exist.'

He is, like Taleb and Wilmott, a mathematician. But their mathematics is not that of the quants. De Vany applies maths to the real world. He wrote the definitive work on the economics of the movie industry which showed that the relation between production costs and revenue was wildly unpredictable. There is absolutely no reason to believe that spending more on a movie will increase its profitability. This offended against the belief of prevailing investment theory that there was a link between investment and return.

De Vany noted that, ignorant of this truth, movie bosses told themselves a series of consoling stories about why a film fails. 'The stories give them the illusion of control, they reinforce prejudices and biases and they all like to feel important.'

The reality, as Arthur discovered, is that 5 per cent of movies pay for the other 95 per cent and success or failure is almost entirely unpredictable. The best the studios can hope to do is find contractual mechanisms that back success after it happens and thus leverage their profits. In fact, this was what they had with their deals with distributors and cinemas and they worked. The movie industry is what de Vany loves best – 'a complex, adaptive, decentralised system', exactly like the human body. He was able to transfer his economics of the external world into the internal.

Dieticians tend to tell similar stories to movies bosses. They believe in homeostasis – that the human body is a self-regulating system that keeps itself in a continuous stable condition – and the balanced diet. In fact, de Vany points out, we are lazy over-eaters. On the African Savannah, where we seem to have spent most of our evolutionary history, food would have been scarce and our bodies were conditioned to eat as much as possible when it was available. The rest of the time we would be hungry for long periods and we would become as lazy as possible to conserve energy. But, in order to obtain animal fats, the highest value food, we would have to walk long distances and be capable of explosive moments of pursuit and killing or, if it all went wrong, of running away. To aim for homeostasis through regularity – of exercise and diet – is to fight against our evolved natures. The body, like the movie industry, is complex, adaptive and decentralised.

At the heart of de Vany's maths is the work of Vilfredo Pareto, an Italian economist and philosopher who died in 1923 and, at the heart of Pareto's work, was the Power Law. Pareto looked at the distribution of wealth through populations. In traditional statistics this would be shown as a bell curve. The high point of the bell would represent the majority of the population clustered around the average wealth figure. The sides of the bell would tail off to include, on one side, the very rich and, on the other, the very poor. But imagine a football stadium in which most of the likely wealth levels are represented. Then imagine that Warren Buffett enters. In all likelihood, he would have more money than all the other people combined. The average would shift sharply upwards but this would be nothing to do with the wealth of anybody except Buffett. The stadium's bell curve of wealth would effectively become meaningless as would the idea of the average. Buffett is rare because his wealth is so extreme and it is that link between rarity and extremity that power laws capture. Essentially, the rarer an event the more extreme it will be.

My first meeting with Nassim Nicholas Taleb in Newport Beach took place while he was spending a couple of days there talking to a conference of bond traders. I sat with him in the hotel café, we ordered tea and I turned on my two voice recorders. He was very agitated, reading every incoming email on his BlackBerry because the Indian consulate in New York had held on to his passport and he needed to fly to Bermuda. People were being mobilised in New York and, for some reason, France, to get the passport. In his agitation, he knocked over his cup of Earl Grey and drowned one of my recorders. I snatched the other away just in time.

The next day he took me to Circuit City to buy two Olympus voice recorders, one for me and one for him. The one for him was to record his lectures – he charges about $60,000 for speaking engagements, so the $100 recorder was probably worth it. The one for me was because he has a high sense of obligation. 'I owe you,' he kept saying. I said it didn't matter because I always use two recorders and, anyway, I had bought a replacement that morning. But it made no difference: he owed me. Happily, the recorders cost $20 less than the marked price owing to a labelling screw-up at Circuit City.

There was a message in this chaos. Stuff happens. The world is random, intrinsically unknowable. 'You will never,' Taleb says, 'be able to control randomness.' The lost passport, the spilt tea were black swans, bad birds that are always lurking just out of sight to catch you unawares and wreck your plans. Sometimes, however, they are good birds – the unexpectedly cheap recorders.

Taleb made his money as a quant, trader and analyst of risk. He was born in 1960 in Lebanon. He calls himself a Levantine, a member of the indecipherably complex eastern Mediterranean civilisation. Both maternal and paternal antecedents were grand, privileged and politically prominent. They are also Christian – Greek Orthodox. He is still a practising Christian, thought not a believer. Religion, he says, is not about belief, it is about custom and practice, about ways of coping with an unknowable and uncertain world.

He was educated at a French school so three traditions formed him – Greek Orthodox, French Catholic and Arab – and they also taught him to disbelieve conventional wisdom. Each tradition had an utterly different history of the Crusades. He learned about the dangers of competing narratives.

But, crucially, he also learned as a child that grown-ups have a poor grasp of probability. This occurred in the midst of the Lebanese civil war when, hiding from the guns and bombs, he heard adults repeatedly say the war would soon be over. It lasted fifteen years. He became intrigued by probability and, after a degree in management from the Wharton Business School at the University of Pennsylvania, he focused on probability for his Ph.D. at the University of Paris.

For the non-mathematician probability is an indecipherably difficult field. But Taleb makes it easy by proving the limitations of a mathematical approach to the real world. He tells the parable of Brooklyn-born Fat Tony and academically inclined Dr John. You toss a coin forty times and it comes up heads every time. What is the chance of it coming up heads the forty-first time? Dr John gives the answer drummed into the heads of every statistics student – 50–50. Fat Tony shakes his head and says the chances are no more than 1 per cent because the coin 'gotta be loaded'.

The chances of a coin coming up heads forty-one times are so small as to be effectively impossible. It is far more likely that somebody is cheating. Fat Tony wins. Dr John loses. Dr John is the economist or banker who thinks he can manage risk through mathematics. Fat Tony is the trader who relies only on what happens in the real world.

In 1985 Taleb discovered how he could play Fat Tony in the markets. France, Germany, Japan, Britain and America signed an agreement to push down the value of the dollar. At the time, Taleb was working as an options trader at a French bank. He held options that had cost him almost nothing that bet on the dollar's decline. Suddenly they were worth a fortune. He started buying 'out of the money' options. He had realised that when markets rise they tend to rise by small amounts but, when they fall, they fall a long way.

The big payoff came on 19 October 1987 – Black Monday. It was the biggest market drop in modern history.

'That had vastly more influence on my thought than any other event in history.'

Nobody had expected it, not even Taleb. But that is the point; the event was unknowable in advance. Almost all events are. The trick is not to try and know the future, but to put yourself in a position in which you can survive it. Taleb was ready for *something* to happen. He had a pile of out-of-the-money Eurodollar options. So, while others were considering suicide, Taleb was sitting on profits of $35–$40 million. He had what he calls his 'fuck off money', money that would allow him to walk away from any job.

He stayed on Wall Street until he got bored and he moved to Chicago to become a trader in the pit, the open outcry market run by the most sceptical people in the world, all Fat Tonys. This, he understood. True market traders react to and respect the complexity of the real world; they do not expect to control it.

All his ideas coalesce around the image of the black swan. Until 1697, everybody in Europe thought all swans were white. It was a simple truth about the world, as obvious as the sun rising in the east and setting in the west. But, in that year, a Dutch captain was exploring the Swan River in Western Australia and he saw a large black bird that was, indubitably, a swan. There are two points about this bird: first, nobody could have predicted its existence in advance and, secondly, it disproved an idea – that all swans are white – that had previously been thought to be beyond all doubt. A black swan, for Taleb, is the emblem of the future we cannot know. It may be good or it may be bad, but, either way, it is entirely unpredictable.

His book, *The Black Swan: The Impact of the Highly Improbable*, came out in 2007 and effectively forecast the crash of the following year. Its central point is that we have created a world we don't understand. There is a place Taleb calls Mediocristan. Here most events happen within a narrow range of probabilities – within the bell curve distribution still taught to statistics students. But we have gradually been leaving Mediocristan. We live in what he calls Extremistan where black swans proliferate and their effect is amplified by modern phenomena such as the world banking system which is highly inter-connected. It is also over-efficient and, therefore, fragile because it lacks any redundancy to protect it against the impact of black swans.

'Every bank,' he said, 'becomes the same bank so they can all go bust together.'

He points out that banks make money from two sources. They take interest on our current accounts and charge us for services. This is easy, safe money. But they also take risks, big risks, with the whole panoply of loans, mortgages, derivates and any other weird device they can dream up. This is an intrinsically fragile business model that must always be on the brink of catastrophic failure.

'Banks have never made a penny out of their investments,' said Taleb, 'not a penny. They do well for a while and then lose it all in a big crash.'

On top of that, he has shown that increased economic concentration has raised our vulnerability to natural disasters. The Kobe earthquake of 1995 cost a lot more than the Tokyo earthquake of 1923. Power laws warn us that the more rare the event, the more devastating its impact.

The world according to Taleb is fundamentally unknowable. We swim in a sea of uncertainty and the attempt to control this world with mathematics or any other system is doomed to failure. Black swans will always expose your illusions. Peace will only be found by acceptance of this. Or, as Taleb put it, 'The only way you can say fuck you to fate is by saying it's not going to affect how I live. So if somebody puts you to death, make sure you shave.'

It is also important to make sure you shave at the right speed. In the early 1990s Paul Wilmott was one of a group of Oxford maths modellers returning from a conference in Helsinki. There was a Gillette advertisement for a twin-blade razor running on TV at the time. This claimed that twin blades gave you a closer shave because the first blade cut the bristle and the second cut it further because it caught the bristle before it could retract, a phenomenon known as hysteresis – basically the lag in a body's response to an external force.

For some reason, while watching one of the air hostesses, Wilmott had the idea that the group could pass the time on the flight by putting together a mathematical model of shaving with a two-blade razor. This might prove to be a spoof, but they could get it published and, in academia, it is a case of publish or perish.

By the time they had landed the model was all but complete and, soon afterwards, it was published in an academic journal. It was a triumph because it showed that there was an optimum shaving speed: too fast and the two-blade effect was lost, too slow and the bristle would retract, evading the pass of the second blade. Somewhere in the Oxford group's equations there was a clear indication that an optimum speed could be established.

Unfortunately, there was a variable in the equations which could not be known. This was the elasticity of the skin which, obviously, varies from person to person. (Wilmott did suggest, less than seriously, that this could be solved using a pair of tweezers and a stopwatch.) Without this variable the model was incomplete, but it could be completed by what is known as an inverse calculation. A shaver could time his shaving and, by trying a range of speeds, arrive at an optimum. This feeding of real-world experiments into a mathematical model is known as calibration.

Calibration can only be relied upon if the underlying model is good, otherwise the calibration will only result in the model generating misleading results. But, under normal circumstances, it will not be obvious that they are misleading because the model will continue to be consistent. Under extreme circumstances the faults in the model will suddenly become apparent. This is what happened in 2008. The crash was implicit in the shaving model.

'You don't find out the model is wrong,' says Wilmott, 'unless you lose a lot of money.'

And, in Wilmott's view, all financial models are, in varying degrees, wrong.

A couple of months before he told me this story, he had been asked to talk to some bank regulators in Washington. Typically, he considered in advance the nature of his audience. Regulators are paid a lot less than bankers and tend, therefore, to be less smart. Wilmott thought of two good, simple questions the regulators could ask the bankers.

'Do you calibrate?'

If the answer was yes, and it always would be, the next question was: 'Do you test the stability of your calibrations?'

At this point, says Wilmott, the bankers would be in a state of shock, realising the regulators were on to them. Calibrating bad models increases risk because it means the faults in the model are concealed by the calibrations. What the banks should be doing is not calibrating but fixing their models.

Wilmott's advice shocked the regulators. One of the rules they imposed on bankers was that they *must* calibrate. They were so shocked, they did not take his advice. The bankers continued to calibrate bad models, their faith in their bad maths and the machine undimmed. The crash should have been a warning to them and can still be a warning to us – there are no simple solutions in a complex world.

NOW NEUROAESTHETICS

What cannot be owned and simplified by the machine? Aesthetics – the study of beauty – has always been owned, though seldom simplified, by the philosophers. Who else could be involved in the attempt to analyse, define and categorise something as complex, intimate and universal as the creation and perception of art? Neither religion nor science was ever likely to have much to say on the subject; only philosophy could be expected to dissect the beautiful.

The philosophers' 2,500-year-old debate oscillates between two primary possibilities. Either beauty is out there in the world and things that are beautiful actually possess beauty within themselves; or it is in here, in our heads, and we make things beautiful with our imaginations. Plato said it was out there: our art was an attempt to recreate the perfect forms of beauty which existed beyond our immediate perceptions. David Hume said it was in here: the beauty of things exists 'in the mind that contemplates them'.

Neither position seems complete. The out-thereists raise the questions: where are these ideal forms and why, if they are indeed ideal, are there so many different types of beauty? The in-hereists, meanwhile, evade the issue of why the sense of beauty itself, rather than its specific manifestations, is so universal and consistent through human societies.

But now the subject is being wrenched from the grasp of the philosophers by the scientists with a new discipline called neuroaesthetics. Their answer to the ancient question? Beauty is mainly Hume – in here – but with a surprisingly large dash of Plato – out there.

Historically, the problem for the scientists was that there were no ways of observing and measuring the creation and appreciation of art and beauty other than through subjective reports and critical analysis, both of which were too laden with personal and cultural overtones to be of any empirical use. There are thousands of reasons for and ways of loving Mozart or Shakespeare, all equally valid and equally resistant to scientific analysis.

This all changed with brain scanners, primarily fMRI machines, which gave clear pictures of activity within the brain. These machines allow us to see what happens when people are making or experiencing the beautiful.

Just as promising for science was the fact that this scanning can be done with the aid of a theory. Neo-Darwinism – the development of Charles Darwin's ideas in the latter part of the twentieth century – offers the possibility of a scientific theory of beauty. Since, to the neo-Darwinians, almost all things in the living world are adaptive responses to the problem of survival, art and beauty are likely to be explicable in terms of some aspect of the processes of natural selection. And so, armed with scanners and a theory, science now advances on one of the most vital and difficult areas of human experience.

There are two sides to this – the creator and the audience. The creator makes beauty, the audience experiences it. Neuroscience pursues not just the artist but also the audience as well as the more general human gift of creativity. Creativity does not necessarily involve painting a picture or writing a poem; it may be a matter of solving a problem in a new way, like, say, finding a new route to work. We all do it many times a day. It is related to artistic creativity – both the artist and the problem solver bring something new into the world and both are now the subject of the most intensive neuroscientific research – but it is not quite the same thing.

In 2002 the discipline of neuroaesthetics was born. For Semir Zeki, professor of neuroaesthetics at University College London and one of the founders of the discipline, the point of the project is more than simply discovering the reactions of the brain to the beautiful. It is also to identify what is distinctive about the human mind. 'There was a tremendous gap,' he says; 'one of the major professed aims of neuroscience is to understand what characterises us as humans – what is uniquely human and what is especially well developed in humans.

'You cannot say decision-making is unique to humans because animals make decisions. But you can say that language is unique to humans, you can say that foresight and hindsight is unique to humans. You can also say creativity is largely unique to humans. Of course, birds build nests and monkeys use tools but there is a kind of creativity that is unique to humans.'

Explaining beauty has always been a challenge to the scientific imagination. Most human behaviours seem to have a clear purpose that can be related by hard Darwinians, however distantly, to basic survival mechanisms like eating and mating. It has always been very difficult to tell such stories about beauty. It is a paradox – a human trait that is both devoid of practical purpose and essential.

'Beauty has no obvious use,' wrote Sigmund Freud, 'nor is there any clear cultural necessity for it. Yet civilisation cannot do without it.'

At the heart of civilisation and at the centre of the human mind lies art. Like religion and love, especially romantic love, which seems at least partially disconnected from procreative lust, art and beauty obsess us. All the evidence we have suggests that as long as there have been humans, there has been art. But why *should* something so apparently useless so preoccupy our species? Freud's solution was to see art in terms of the inevitable conflict between the instinctive violent and sexual urges of the biological self and the demands and restrictions of civilisation. Frustrated, we seek alternative satisfactions, one type of which Freud calls 'substitutive'. 'Art,' he wrote, 'is a conventionally accepted reality in which, thanks to artistic illusion, symbols and substitutes are able to provoke real emotions.'

All substitutive satisfactions are 'illusions . . . but they are nonetheless psychically effective, thanks to the role which phantasy has assumed in mental life'. Thwarted by society in the pursuit of our most basic urges, we redirect our psychic energy to the pursuit of beauty.

This story may be convincing in its own terms but it is not good enough for the neuroscientist. It may – just about – fit the Darwinian model but, as an explanation, it works at too high a level. Freud's is a story about how art might come about, not about what it actually is. Discovering this is the goal of Semir Zeki.

He is a small, bald seventy-year-old with a gleaming, Yoda-ish face. When we meet he is wearing a very light blue and white Japanese jacket over his pullover which I at first take for some unusually short and colourful academic gown. His room, like every other room I have ever seen in UCL, is a mess – piles of books, furniture near collapse and all illuminated by the weak glow of a pale, greyish sunlight of a type that I have only ever seen in this place.

He is eager and brisk, his accent is upper-class English but not quite. His eagerness is that of the scientist who knows he is at the cutting edge – at last, we can watch beauty, the soul of humanity, as it happens.

'Things have changed,' he says, 'to such an extent that now I can actually objectively measure a subjective mental state. In other words if I were to show you twenty paintings and half of them you would declare to be ten out of ten in beauty as you see it and the other ten are ugly as you see it, then I can find activity in a given part of your brain and I would know this and I can see the ones you declared to be beautiful would show greater activity than the ones you declared to be ugly.'

'So you would know if I was lying?'

'Oh, indeed, I would know you were lying, yes.'

He is a man who lives in art. He is spoken of as the scientist who wanders around listening repeatedly to Bach's Goldberg Variations, a series of short, enigmatic piano pieces that seem specially composed to inspire speculation and wonder.

'Variation number twenty-five,' he says with a certain satisfaction, 'to me that one is ambiguous and recently I have been able to interpret it in the following way – it is the question "What is existence?" repeated and repeated and repeated and never answered.'

He also tries to go to every performance of Wagner's opera *Tristan und Isolde*. But, such is his immersion in the opera, he is invariably disappointed. No performance is ever good enough.

'I am very irritated by the performances because they depart too much from my concept of the ideal performance.'

Zeki is a scientific maverick. He does not just write about brain scans and anatomy, he also writes extensively about art, psychology and philosophy, the humanities in general. This inspires mistrust in other neuroscientists. He is felt to have gone off-message by failing to stick to hard science, but he is unapologetic. For him, it is hard science that is off-message.

'Certain ideas,' he says, 'have not been accessible to science for the simple reason that scientists tend to become rather arrogant. A rather well-meaning young man once said to me after a lecture that what I was saying was not science because science is about measurement. I said to him science is not about measurement, it is about curiosity and measurement is a means of satisfying that curiosity because it carries with it the weight of predictability and confirmation. That's all.

'What they call hard science is this, the measurement of transmitters in the brain, the measurement of reaction of single cells is hard science. Soft science is when you ask questions about the love system of the brain or when you say you've got some research from literature of the love system of the brain.'

He has just had a paper rejected in which he had shown that romantic love in gays and lesbians was no different from heterosexual love. The editors of the journal had objected to a paragraph in which he quoted Michelangelo and the Persian poet Rumi. Such things, they said, had no place in a scientific paper and, anyway, the subject was not of general interest. But, for him, it is simply irrational for science to ignore the evidence of the humanities.

'When you consider these factors and also the strong religious side of humans, which I believe to be linked to the love system in the brain, then you have to go back to the stone age if you want to do it without the humanities because these issues have been discussed in the humanities from Plato onwards. Neurobiology has invaded everything but the one thing it has not invaded is the humanities and I think they have a tremendous amount to offer.'

Zeki was born in Beirut to a father of Lebanese descent and a mother who was – he waves his hand vaguely – 'more European'. His father's work bought him to Britain where he studied anthropology and then swerved into medicine, which he also discovered was not for him.

'I realised that in medicine once you have seen the first hundred patients, you've seen them all, and, if you specialise, once you've seen the first fifty patients, you've seen them all.'

There are two messages about the man here. He is easily bored and he likes to push against the boundaries of any discipline.

At UCL his interests focused on the brain, specifically the visual brain. But, again, he felt the need to push the boundaries and so, in the late 1990s, he invented a new subject – neuroaesthetics.

His view of our sense of beauty is dominated by two principles derived from neurobiology – constancy and abstraction, both of which came from his study of the visual sense. Constancy is fundamental to the way we make sense of the world. We know a chair is the same chair even when we see it from different angles or in different lights. Abstraction is the way we classify things in the world. We do not see just this chair; rather, we know it is a chair because it belongs to a group of many different things which are also chairs. The puzzle of how, exactly, we do this has dogged Western philosophy. Zeki has gone back to the roots of the debate by agreeing, at least partially, with Plato. Like Plato, he believes that the class of objects we call chairs has real existence. But Plato would have said the class exists outside the human mind. Zeki agrees that the class really exists but inside the mind. Our minds are built to construct abstractions which we use to organise the world.

This fits neatly within the Darwinian framework. If a lion had killed my brother, then it would be useful if I recognised lions in general rather than just that particular predator so that I could take evasive action. It would also be useful if I knew that a lion in the light of dawn was still a lion in the light of dusk. It is clear, therefore, that constancy and abstraction are fundamental to our survival.

The creation of ideals – classes – in the mind is central to Zeki's neuroaesthetic view of art – and of religion and love. What is wrong with Freud's view, according to Zeki, is that he talks as if the world and the mind were two quite different things. 'Freud writes as if there are two wholly distinct entities,' he writes, 'the external reality and the internal reality, without acknowledging that there is only one reality, brain reality, which is shaped by both external and internal influences.'

The brain reality is haunted by idealisations. These may have evolved as survival mechanisms but they created an abyss between what was realisable in the world and in the mind. Our ideals, as we all know so well, seldom survive their first contact with reality intact. Art is an attempt to bridge this abyss, to realise the ideal in the world. This is, of course, impossible, which is why, for the artist, art is so often about failure. Michelangelo failed to finish many works; Samuel Beckett wrote in a clear reference to his own artistic struggles, 'Fail again. Fail better'; and Paul Valéry insisted, 'A poem is never finished, only abandoned'.

So where Freud said art was a defence against or distraction from the conflict between the instincts and society, Zeki says it is a gloriously doomed attempt to make something in the world that matches the ideal in the mind. Freud remarked that civilisation cannot do without beauty. Zeki says this is because it is a 'neurobiologically essential'. The pursuit of beauty is the pursuit of peace within ourselves.

Zeki's neuroscientific conception of art is in competition with that of Vilayanur Ramachandran, the most celebrated neuroscientist alive. Ramachandran, born in India in 1951 and now a psychology professor at the University of California at San Diego, is best known to the lay public for his work on phantom limbs.

Amputees often retain sensation – pain, discomfort, itching – in the missing limb, but are, obviously, unable to do anything about it. Ramachandran's insight was that the signals from the limb had been locked into the brain. He devised a mirror system so that the patient 'saw' the missing limb and could move it. Over time, the memory of the limb was removed from the brain. The device was used in a dramatic scene in the US TV series *House* in which Dr House cures a man of agonising pain which seems to come from his amputated arm by creating a mirror box in which he sees the reflection of his good arm as his missing arm.

Ramachandran's less celebrated – perhaps because more complicated and less dramatic – theory of artistic experience is, admittedly, speculative. Like Zeki, he starts from the visual sense, but, instead of two crucial factors, he has eight laws, which include grouping, isolation and symmetry. Zeki is impatient with this level of complication, but Ramachandran is actually attempting to explain something in detail while Zeki is looking for the more general pattern. They do not, in other words, necessarily contradict each other.

The neo-Darwinians have yet another perspective. Music seems to preoccupy them more than the visual arts, perhaps because it is the most abstract of arts and, therefore, the most inexplicable. Music engages us deeply but it can never be clearly stated by what we are engaged.

'What I like about music,' writes John Ashbery, 'is its ability of being convincing, of carrying an argument through successfully to the finish, though the terms of this argument remain unknown quantities.'

Steven Pinker, psychologist and cognitive scientist, has argued that music is an accidental by-product of language. From within the same disciplines, Daniel Levitin disputes this, saying music may be our peacock's tail, an example of runaway sexual selection. Female peacocks like males with a big tail and then a bigger tail. In the same way, in humans, art takes on ever more elaborate and ingenious variations.

Art is, therefore, an intimate interaction between the artist and his audience, as intimate as courtship rituals. The peacock's tail seems to be a good analogy. It is plainly bad for the bird, restricting its movements and requiring an inefficient expenditure of energy. But it has become necessary to attract a mate. This, in humans, would explain the superfluity of art which, Oscar Wilde insisted, is quite useless. But, like the peacock's tail or, indeed, like the human brain, it may simply have become a flamboyant excess driven by, ultimately, the requirements of survival and reproduction. It might also explain the disappointment we feel when we discover a picture is a fake. It is as if the tail on the peacock had turned out to be an artificial contraption and we have been denied contact with the real thing.

Yale professor Paul Bloom has a theory that lies closer to the insights of neuroscience. Bloom is an evolutionary psychologist, which means he uses Darwinian natural selection to explain psychological traits, just as it is normally used to explain the development of physical organisms.

He starts from the observation that we often value objects irrationally. For example, there was a tape measure owned by President Kennedy that sold at auction for $48,875, the shoes thrown at President Bush by an Iraqi journalist sold for $10 million, the seventieth home run baseball hit by Mark McGwire which sold for £3 million. There is nothing about these objects *in themselves* that makes them so valuable. They are only valuable by association.

In some cases, there seems to be a superstitious feeling that the objects themselves are physically changed by the association. A sweater owned by George Clooney, Bloom points out, is valuable but becomes less valuable if it is sterilised, as if this has removed some essence of the star. In a further twist, he points to an experiment in which the violinist Joshua Bell played in a subway. People pay hundreds of dollars to see Bell perform in a concert hall but a thousand people walked past him when he busked and he only made $32. He was still Joshua Bell, the great violinist, but, somehow, in the subway, he was not.

Bloom's theory is that humans are naturally essentialist. We see everything as possessing 'an underlying reality or true nature that one cannot observe directly and it is this hidden nature that really matters'. Show children a toy tiger and they will say it is a tiger. Wrap it in, say, a fox's skin and they will still insist it is a tiger. We also value gold for no practical reason, only because we think it is imbued with some inner quality.

For Bloom, our essentialism is potentially explained by evolutionary psychology. 'At first it seems really weird that we value the history of objects so much. What difference does it make? I think it's a spillover from the fact that the history of people really does matter to us. If you are this person who did this favour for me a year ago or the one that betrayed me – well, it is really important for me to analyse you. This has spilled over into things that people create.'

We are evolved to be instant psychoanalysts and we transfer

the skill from people to things. What we consider art or merely valuable is simply the attribution of an unusually intense form of inwardness to an object or composition.

But then there is the more mundane issue of the creativity of problem solving. Sherlock Holmes, to the mystification of Dr Watson, would take time off from investigating a case to play the violin. He was apparently good at it. Indeed, the Sherlock Holmes Society of London speculates that Arthur Conan Doyle may have named the great detective after Alfred Sherlock, a leading violinist of the time.

Watson was mystified because he had a traditional work ethic – you work doggedly and dutifully until the work is done. But Holmes was a genius and, far from ignoring the case at hand by picking up his fiddle, he was resorting to an extremely efficient way of solving it – by taking his mind elsewhere.

And, to return to the TV series *House*, Dr Gregory House, a character modelled on Holmes, uses the same technique. He diagnoses the most obscure ailments only at the critical moments when he is thinking about something entirely different. Dr Watson and House's team of lesser doctors are uninspired minds who see only distraction when in fact the mind of the genius is, in apparent indolence, working at full stretch.

That is fiction, but there is also fact. Notably, there is a particularly famous case, one of the most studied in the whole neuroscience of creativity. At 6 p.m. on 5 August 1949 a fireman named Wag Dodge along with his crew found themselves cut off by a wildfire in Mann Gulch River Valley, Montana. A wall of flame was coming towards them at 30 mph. Dodge took a match out of his pocket and set fire to the grass immediately in front of him, stepped into the cleared space, covered his face and pressed himself into the ground so that he could breathe the thin layer of air beneath the smoke cloud. The fire rushed over him and he survived. The other thirteen members of the crew hadn't heard his order to do the same. They all died.

'Wag Dodge – he's a great one,' said Mark Jung Beeman, a professor of psychology at Northwestern University in Evanston, Illinois. 'It was particularly interesting in such a stressful situation. He was at the point where he basically gave up. He must have had some pretty awesome frontal lobes. Normally high stress would limit creative, flexible or insight type of thinking but not in this case.'

The key words here are 'where he basically gave up'. Dodge had been struggling to find a way to escape the flames for some time. When, finally, the situation seemed to be hopeless, he had a moment of relaxation, of giving up, and that is when he saw the solution.

Dodge was Holmes playing his violin or House gossiping about sexual intrigues in his office. He switched off from the immediate problem at a critical moment and at once solved it. (In fact, as I shall show, this seems very close to the inspiration of artists.)

Dodge was inspired, but how? The assumption that he breathed the breath of God – the etymological root of the word 'inspiration' – is obviously a dead end for scientific inquiry. In fact, the word itself seems almost to be banned in some circles.

'It's not really inspiration,' says neuroscientist Earl Miller, 'there's really no such thing. It's more like a reconfiguration of old thoughts. I know from my own experience that most of my insight comes when I'm not thinking about a problem. I work until I'm really caught in a rut and then I take a walk or play music or when I drift off to sleep the solution will occur to me.'

Creativity is about the remaking, the discovery or invention of new connections. Neuroscientists are reluctant to call this 'inspiration'; they prefer the word 'insight', which removes the mystical overtones. 'Inspiration' suggests something new has come into the world, something lying beyond the mechanisms of material causality. 'Insight' suggests only that the individual has seen something from a new perspective or in a different form.

But is this just a change in vocabulary? Does inspiration itself remain untouched? The main problem seems to be the sheer skittishness of our brains, their promiscuity in the face of the world.

'We are literally remaking our brains,' writes the neuroscientist Nancy Andreasen, 'who we are and how we think, with all our actions, reactions, perceptions, postures, and positions – every minute of the day and every day of the week and every month and year of our entire lives.'

Every event in your life, everything you see, hear, touch, taste, smell and even think, physically alters your brain. This may seem obvious – something *must* be happening in there – but it is bewildering nonetheless. Within the confines of your skull, from moment to moment, a few pounds of fat and water can both make and remake an entire world. Your brain, before it is anything else, is creative. But how?

All we can say, perhaps, is that disorganisation is a part of the answer. Inspiration or insight, in these terms, seems to come from the moment when connections within the brain are relaxed, when there is 'a dissociated pattern of activity' in the prefrontal cortex – the area of the brain in which the highest cognitive functions take place. Dissociation is the moment of relaxation – the moment Wag Dodge gives up. The 'dissociated pattern' is that it echoes the loosening of connections that precedes the 'Aha!' moment when the problem is solved. Insight and creativity, perhaps even genius, do seem to be linked to a brain that can disorganise itself, freewheel, making new and unexpected connections.

As Nancy Andreasen puts it, the creative act may 'begin with a process during which associative links run wild, creating new connections, many of which might seem strange or implausible. The disorganised mental state may persist for many hours, while words, images and ideas collide. Eventually order emerges and with it the creative product.'

This once again points to the inadequacy of machine models of the mind. If the brain is thought of as a computer or as any recognisable type of machine, then plainly it works differently from anything we have ever made. We do not expect our laptops to be distracted, dissociated or disorganised before they discover how to follow our commands. There is a fundamental difference here between what we now know of the mind and our machines, and, if this difference cannot be explained scientifically, then the human mind in its highest, most creative actions would, once again, be in danger of escaping the clutches of the materialist world view.

But there are a few more clues to this mystery in the science. Everybody is not equally creative, equally able to loosen the connections. Tests have been devised to measure degrees of creativity. One involved asking people how many uses they could imagine for a brick. Such tests have been attacked as far too subjective. But they do point to a crucial way of defining creativity. If you can imagine dozens of uses for a brick, then you are what is known as a divergent – more creative – thinker. If, instead, the question makes you impatient and strengthens your conviction that bricks are for building walls, then you are a convergent thinker – more analytical.

Divergent thinkers habitually wander around their own minds, looking for links, however absurd or surreal. Convergent thinkers look for the one correct answer. So the discovery of the structure of DNA by Watson and Crick in 1953, for example, was a clear example of convergent thinking – the one correct answer was a double helix. The minds of Mozart and Shakespeare were, in contrast, limitlessly divergent.

But, of course, creative divergers who can think of 101 uses for a brick are treading a fine line. Too many divergences can drive you insane. People have always suspected there is a link between madness and genius. As John Dryden observed,

Great wits are sure to madness near allied,
And thin partitions do their bounds divide.

What science we now have suggests that it might be true. Oddly, however, high creativity has not so far been found to be linked with schizophrenia, as most people expected, but with mood disorders – notably bipolar disorder or manic depression.

The link has been made by several highly authoritative studies, all by leading American scientists. Kay Jamison, a clinical psychologist at Johns Hopkins in Baltimore, studied poets, playwrights, novelists, biographers and artists and found 38 per cent had been treated for an affective illness – i.e. mood disorder. Joseph Schildkraut, a Harvard psychiatrist, studied fifteen abstract expressionist painters from the 1950s – 50 per cent had psychiatric issues, mainly mood disorders. And Nancy Andreasen studied students at the Iowa Writers' Workshop, the leading school of its kind in the world. Again, there was a phenomenally high percentage of mood disorders.

The idea of the dissociative, divergent mind that flirts with madness has a strong attraction. It provides a scientific narrative that justifies our most romantic conception of the genius as a disordered outsider whose very eccentricity is evidence of the penetration of his insights. It also, perversely, feels like an anti-scientific narrative in that it seems free of strict rationality, of traceable links between cause and effect. We may talk of the relaxation into dissociated patterns of the prefrontal cortex, but that is no more than an accompaniment to the moment of inspiration, a necessary but not sufficient condition.

The much deeper question is: why is this particular product of the dissociated prefrontal cortex especially significant? Plainly the fourth movement of Mozart's Jupiter Symphony has more meaning for more people than some simple tune that might flash into my mind. We may all be inspired all the time, but very few are inspired in such a way that, in effect, the whole of humanity can share in the inspiration. What happens in the mind of the great artist that makes his work universally significant? To understand this, it is necessary to understand his audience.

Curiously, Darwin, the dominant figure in all such accounts, fell silent before the problem. Or, rather, he resorted to aesthetics as the philosophers, rather than the scientists, understand the word. At the end of *On the Origin of Species*, to rescue some human dignity from the cataclysm of the vision that seemed to say we were just one more animal, he turned to beauty as a saving grace:

> There is grandeur in this view of life, with its several powers, having been originally breathed into a few forms or into one; and that, whilst this planet has gone cycling on according to the fixed law of gravity, from so simple a beginning endless forms most beautiful and most wonderful have been, and are being, evolved.

His terms are purely aesthetic – 'most beautiful and most wonderful' – we should see the workings of nature as beautiful, a great work of art. What he could not do was claim this aesthetic sense could be encompassed by his own theory. So the perception of beauty does, in its way, offer some hope to the idea that humans are exceptional, not just another species locked inside the logic of natural selection. We, after all, dreamed up the theory and we could see the beauty.

This degree of human exceptionalism persists, if sometimes weakly, among contemporary neo-Darwinians as a celebration not of beauty but of reason. 'We, alone on earth,' Richard Dawkins has written, 'can rebel against the tyranny of the selfish replicators.' And Steven Pinker famously remarked that, if his genes don't like what he does, 'they can go jump in the lake'. A human, Pinker said, was 'simultaneously a machine and a sentient free agent'. The implication here is that Darwinism cannot fully explain humans. Our peacock's tail – the mind – has generated a rationality that can reject the crude demands of survival and reproduction. It has also generated art and beauty.

The puzzle remains. How does the new, the dazzling insight,

the absolutely fresh perspective, emerge from the mind?

This is not merely a matter for philosophers and scientists. At a less exalted level than art, beauty and neuroscience, this question has penetrated the popular imagination in the language of management theory.

'Thinking outside the box' – meaning thinking originally – or 'lateral thinking' have long been pressed upon employees and managers as ways of generating new ideas. Latterly these have been superseded by the concept of 'vuja de', a mangling of the French expression *déjà vu*. The latter means something already seen or known; the former denotes something you may have seen or known but you feel you are encountering it for the first time, you are seeing the world anew. This leads – or, rather, inspires – you to ask unusually basic questions like, why do people wear a watch when they have a mobile phone that tells the time? The 'vuja de' trick is to jolt you out of habitual modes of thinking. It is intended to be a way of, as the neuroscientists would say, causing your brain to dissociate and pave the way for inspiration.

Such ideas had been nascent in Western culture thanks to the popularisation of Eastern modes of thought, notably Zen Buddhism, from the late 1950s onwards. Zen insight wisdom proceeds through surprises intended to produce new states of mind. So the *koan* – a nonsensical question like 'What is the sound of one hand clapping?' – is intended not so much to be answered as to change the terms of your perception of reality.

Zen's appeal in the West may be something to do with the possibility of escape from too rigid modes of thought and discourse. But it also offered a serene sense of acceptance of the power of the human imagination. The most effective evangelist of Zen in America, Shunryu Suzuki, captured in a sentence its appeal to the mind that yearns for inspiration: 'In the beginner's mind there are many possibilities but in the expert's mind there are few.'

This suggests a great truth about the artist – he is the absolute beginner, the creator of the new.

THE NEW FOUND LAND

> Its vanished trees, the trees that had made way for Gatsby's
> house, had once pandered in whispers to the last and greatest
> of all human dreams; for a transitory enchanted moment man
> must have held his breath in the presence of this continent,
> compelled into an aesthetic contemplation he neither under-
> stood nor desired, face to face for the last time in history
> with something commensurate to his capacity for wonder.

At the end of *The Great Gatsby*, F. Scott Fitzgerald's narrator, Nick
Carraway, has a vision of the past, of the fading of the human
presence on Long Island Sound. He sees the land return to a still
undiscovered wilderness and, as the trees reclaim their territory,
Nick imagines the feelings of the first Dutch sailors to arrive on
the eastern shore of this new continent. The sailors, in his vision,
become a singular 'man' – Everyman – 'face to face for the last
time in human history with something commensurate with his
capacity for wonder'. For this one last moment of discovery, man,
neither understanding nor desiring, was forced into 'an aesthetic
contemplation'. This new continent, America, was big enough
and strange enough to fill his soul. He became a beginner again,
open to the possibility of the new.

This passage is about a moment of transformation, for Nick
and for the sailors. It is a moment pursued but not achieved by
Nick's doomed friend, Jay Gatsby, though, in death and in art,
Gatsby is included in the wonder of the new continent. The
transformation is a shift in perspective. The sailors suddenly see

the limitations of their old world and Nick sees Gatsby's pursuit of Daisy, his lost love, in terms of the aesthetic contemplation of a new world. For Fitzgerald, as for the European explorers, America is, above all, a chance to 'start over'.

But the real subject of the passage is art. Nick's vision is the moment of inspiration, the moment at which everything falls into place and the stories of Gatsby and of America are seen as if for the first time. The wonder of Nick and the sailors is innocent. They are shocked into innocence by the spectacle of novelty. Something absolutely new – America, the truth about Gatsby's love – appears in the world and, in this moment of discovery, the human mind is forced to become that of a beginner possessed, in Shunryu Suzuki's phrase, of 'many possibilities'.

Art is discovery. The very word 'discovery' suggests parallels with the scientific method. Einstein glimpsing relativity or Galileo seeing the moons of Jupiter may well have found themselves with beginners' minds, just like Nick Carraway. So are art and science ultimately one?

This art–science link at the moment of inspiration was brilliantly documented in the 1940s by the French mathematician Jacques Hadamard. He wrote to scientists, mathematicians and artists to ask them how they worked. Hadamard had been startled into this research by his own thought processes while trying to solve mathematical problems. He noticed two curious and unexpected aspects of his method. First, he found he did not think in words or even numbers. Instead, he saw mental images that suggested solutions. Secondly, he noticed that days of struggling with a problem often produced nothing. But then, when he stopped struggling and while he was doing something else entirely, the solution would suddenly flash into his mind.

Many of his correspondents reported similar experiences. Albert Einstein spoke of 'certain signs and more or less clear images', and others of solutions emerging, not from patient and logical reasoning, but from nothing and nowhere when the mind was otherwise engaged.

Hadamard made the link to artistic inspiration via a letter written by Mozart describing his creative method. 'When I feel well,' he wrote, 'and in a good humour, or when I am taking a drive or walking after a good meal, or in the night when I cannot sleep, thoughts crowd into my mind as easily as you could wish. Whence and how do they come? I do not know and I have nothing to do with it. Those which please me I keep in my head and hum them; at least others have told me that I do so ... Then my soul is on fire with inspiration.'

The question 'Whence and how do they come?' implies that the work in some way existed prior to its appearance in the composer's mind, an intuition experienced by many artists. It is as if they are not the creators but the receivers of the work.

Mozart's letter then goes on to describe a creative process that closely matches Hadamard's own timeline of scientific creation: preparation, incubation, illumination and verification. At the point of illumination, the composer could see the whole work in his head. His version of 'verification' consisted of actually writing the score.

Aaron Sorkin, the writer of the TV series *The West Wing* and the movie *The Social Network*, the story of the beginning of Facebook, described something very similar to me. First comes the research, in Hadamard's terms the preparation. 'You feel good during that,' Sorkin said, 'because there's a reason why you're not writing: you can't.'

Then comes the incubation: 'Climbing the walls, pacing around, driving around in my car, making my friends and family miserable and I shower a lot.' Then there is the illumination, a phase in which he writes in a kind of paroxysm which seems unstoppable – he writes some of the longest scripts ever seen in Hollywood. Verification happens on set when the script is tested in performance.

Sorkin has also described the agony of actually starting to write. 'The difference between being on page two and being on page nothing is the difference between life and death. I can't stare at that blank page with the blinking cursor; it drives me mad. I want my foot in the door, I want to get started. And, once I've got started, I want to get to the end, and once I'm at the end, I know so much more about what it is that I'm writing that I can go back and take out everything that isn't about what I was writing.'

Throughout the whole process, it is as if he is waiting for the script to happen to him, as if it already exists, but, somehow, he has to find it. This echoes the fact that, for Sorkin, his art emerged as a discovery, a means of escape from a feeling of inadequacy. As a child, he felt intellectually inferior to his family and friends. He often could not follow their conversations and, as a result, he heard them as almost abstract sound forms. Taken to the theatre to see plays too adult for his understanding, he was entranced by the sound of the dialogue. 'I didn't really understand what was happening up there but I loved the sound of the dialogue, it sounded like music to me and I wanted to make that sound.'

Art, for Sorkin, is a double discovery – first, he discovers that there is a form of art that reflects his childish fascination with the sound of spoken words, secondly, his script writing is an act of discovery of something that seems already to exist.

This sense of the pre-existence of the work is mysterious but commonplace. Bob Dylan notes the same phenomenon. 'The songs are there. They exist all by themselves just waiting for someone to write them down. I just put them down on paper.'

Rumi, the thirteenth-century Persian poet and Sufi mystic, celebrated the act of allowing the work to happen to the artist in a moment when the mind relaxes. He saw this as a moment of surrender:

So let us rather not be sure of anything,
Beside ourselves, and only that, so

Miraculous beings come running to help.
Crazed, lying in a zero circle, mute,
We shall be saying finally,
With tremendous eloquence, Lead us.
When we have totally surrendered to that beauty,
We shall be a mighty kindness.

And William Wordsworth celebrated the times when the mind relaxes, opening the gates of memory and inspiration:

For oft, when on my couch I lie
In vacant or in pensive mood,
They flash upon that inward eye
Which is the bliss of solitude.

At such moments, it seems, science and art merge. Both the scientist and the artist are struck by a moment of inspiration – the 'flash upon that inward eye' – that seems to come from nowhere and to happen to them rather than being achieved by deliberate effort. The new thing pre-exists but not, perhaps, in this world.

But there is a difference, one that sets the two realms apart and asserts the autonomy and importance of art against the claims of science.

Science accumulates its wisdom and ideas are routinely invalidated by later insights. Einstein may or may not have been a greater genius than Newton but his physics is better – in the sense of being a more accurate picture of the universe. Darwin's biology is primitive compared to our present understanding of the living world. But a painting by Rothko cannot be said to be better than one by Titian. Picasso, admittedly, may be a greater painter than Monet, not because he comes after Monet but because he is, indeed, a great painter judged by standards that have nothing to do with the times in which they lived.

Because art does not accumulate in the same way as science, it would be absurd – and demonstrably false – to claim that composers who came after Mozart wrote better music than Mozart just because they were later. Einstein may invalidate Newton but how could the Jupiter Symphony be invalidated? The beginner's mind is born again and again every time it is heard. This does not happen with every new observation that the planet Jupiter has moons. Science departs, but art is in a constant state of arrival. It is always new.

'Literature,' wrote Ezra Pound, 'is news that stays news.'

Perhaps fearing that art may escape the claims of materialism, some scientists resist this distinction. I recently found myself in a very heated argument with the physicist David Deutsch because of his insistence that art should embrace progress in the form of the cumulative approach of science so that later art would tend to be better than earlier. Deutsch used painting as an example. The gulf between us was unbridgeable as he could not imagine any higher virtue than progress and I found his conception of art incomprehensible.

An aspect of Deutsch's argument was the fact that artists always strive to be better than previous artists and therefore, in his terms, they desire progress. This is true and most clearly shown by the way artists, especially visual artists, rush to embrace new technology. David Hockney, one of the greatest draughtsmen alive, now draws exclusively on an iPad. Video art in the hands of fine artists like Bill Viola now successfully competes with paint. The movie industry, meanwhile, is using technology to increase the impact of films seen on the big screen as opposed to on televisions or computers.

James Cameron, for example, first conceived *Avatar* in the 1990s but put off actually making it until the technology could catch up. He was waiting for two crucial developments: 'performance capture' and 3D. Performance capture meant he could load the actors with sensors, film them and then convert them in a computer into the tall blue aliens, the Na'vi, his script required. Films had previously been made in 3D, of course, but the effect had never risen above the level of a gimmick. New digital technology meant it could become much more realistic.

'You have to deal with these technologies separately,' Cameron told me, 'performance capture is not something all film makers are going to embrace. Not every film requires characters that are humanoid enough to be played by actors and inhumanoid enough not to be done by make-up. If you're doing a talking mouse, you're not going to use that technique and if you are doing something closer to human then you'll use make-up.

'It was fundamental to making *Avatar* because it created the sense of visceral otherness in the characters and yet animated the human emotional affect.'

The success of *Avatar*'s 3D is now likely to change the movie business.

'You have a legitimate established market for 3D now and the only question is: is it going to be as ubiquitous as colour? I made a decision five or six years to make all my future films in 3D.'

But it is doubtful that even the supremely confident Cameron would claim that these technologies meant that these movies were greater art than anything that had gone before. Apart from anything else, in the hands of another director *Avatar* could have been a complete mess, unredeemed by the technology. That means, of course, that we judge art by standards other than technological development and scientific models of progress.

But artists do need to 'make it new', as Ezra Pound put it, because newness, as Fitzgerald so movingly demonstrated, is in the very nature of the artistic experience. Or, to put it another way, if *Avatar* is a great film, then it will still be a great film in a hundred years time when its technology will seem primitive.

The wonder of art is, in fact, a wonder based on the impossibility of progress because the best of what is old is perpetually new – it is 'news that stays news'. None of which is to diminish scientific wonder, but it is to say it is different from artistic wonder. Scientific wonder may be a constant through the centuries, but the occasion of wonder – the moons, relativity – is always changing. With a work of art, the occasion is always the same. This suggests, to me at least, that art is a fundamentally different form of knowing from science; it is not new knowledge *about* reality, it is, in Wallace Stevens' words 'a new knowledge *of* reality'. What we know through art does not concern the parts or systems of the world; it does not even concern the whole of the world. Rather, it concerns the world and our place in it. Art is *necessarily* complex and *necessarily* new even when – or especially when – it is old.

This makes the source of the artist's inspiration all the more mysterious, as the thing itself – the revelatory moment that includes everything – is not, by definition, reducible, explicable in terms other than itself. It can be mystifying and frustrating to the mind of a genius that many people do not understand this and demand explanations that attempt to reduce the thing itself to smaller, more manageable parts.

'You cannot complain,' said Samuel Beckett of James Joyce's novel *Finnegans Wake*, 'that this stuff is not written in English. It is not written at all. It is not to be read. It is to be looked at and listened to. His writing is not about something. It is that something itself.'

The same is true of that paragraph of Fitzgerald's. Much can be said about it, it can be paraphrased or explained, but nothing can *become* it.

Perhaps the attempt to equate art and madness is a quasi-scientific attempt to contain and control the unique phenomenon of art. Certainly, it has been tried with Mozart, who has a good claim to be the most bewilderingly inexplicable and profligate of all artists. He is said to be one of many great creators – W. B. Yeats, Beethoven, Melville, Kant, Warhol, van Gogh – who suffered from Asperger's syndrome, a form of autism. But does this explain anything at all? Asperger's is 93 per cent heritable, but great artists seldom produce similarly gifted offspring. In addition, most great artists do not have Asperger's. So the diagnosis, even if true, only seems minimally correlated with exceptional creativity. In any case, isn't there a danger that we are simply grouping together exceptional behaviour and then taming it with a name?

The reality is that our accounts of art are as partial, unformed and inconclusive as our accounts of the human mind. This is perhaps because they are accounts of the same thing. Just as, in Nancy Andreasen's words, 'we are literally remaking our brains' all the time, making them anew, so art literally remakes reality and the artist is the person who possesses – and who offers us all – the beginner's mind. This is another way of showing the inadequacy of the machine model of the human mind.

But this newness is a dangerous quality. First, it is easily mis-understood. As T. S. Eliot pointed out, the wholly original would be wholly incomprehensible so novelty must be defined by what has gone before. The beginner's mind must also be, like the artist's or scientist's mind at the moment of inspiration, the mind prepared by contact with the past. Secondly, as emerged in my argument with Deutsch, creative novelty is dangerous for the prevailing materialist account of the world and the mind. If reality is a chain of cause and effect going back all the way to the Big Bang and, for all we know, beyond, then newness is an impossibility, everything that happens must, somehow, be embedded in what previously happened. The Jupiter Symphony is as inevitable as the moons of Jupiter.

Artists are usually aware of the dangers of what they do and many pursue danger as an end in itself. Jackson Pollock risked an excess of originality when he dripped paint on his canvases, at which point he was, as John Ashbery has pointed out, either the greatest painter in the world or he was nothing. But, also, just as dramatically, there is the British artist Michael Landy.

He is an artist with an innocent streak and eyes that seem permanently fixed in an expression of wonder. In 2001 he took over a disused clothes stores in London's Oxford Street for his work *Break Down*. He set up four work bays, various machines and conveyor belts and spent two weeks shredding and granulating everything he owned, a total of 7,006 objects including valuable art works, his car, his passport and tax documents. Destroying the latter two was illegal but the authorities let him off.

'I'm ruthless,' he said, 'I had got to be ruthless. In the end I had to adhere to the idea, the structure. I'd have felt a terrible sense of regret if I hadn't actually destroyed everything.'

Visitors were enthralled, even jealous.

'It must,' said one to me, 'be very cleansing.'

Sometimes, he felt he was attending his own public execution or funeral with friends, acquaintances and strangers turning up to watch and mourn. On the last day his mother arrived and burst into tears.

'I had to be really tough with her and say, "Look, Mum, you've got to go." She was worried about me, I guess. I didn't want any kind of public displays. I couldn't handle it basically.'

Landy did not want visitors to look at individual items because, he said, 'everybody has more or less the same things'.

Landy's absolute commitment – even the shredded remains of his possessions were to be lost as landfill – gave him a Christ-like quality. Nuns turned up and vicars contemplated the process to acquire material for their sermons. The more secular minded also wanted to get in on the act. A psychiatrist concluded Landy needed help and suggested he get in touch once the show was over. This is machine-thinking. In reality, Landy had become a kind of Holy Fool, a truth-teller. Only to the less imaginative mind was he merely deranged. One visitor to *Break Down* put a note on his conveyor. It said 'Save Me'.

At one level, the work was plainly about the ephemerality of the fruits of consumer society. But, at a much deeper level, it was concerned not with what was destroyed but what was left at the end of the process. The answer was Landy himself, but Landy remade. We flee into possessions, extending our selves into them, like the prosthetic iPod or smartphone. But if we remove these extensions we are driven back into our bodies. Landy had forced his mind to be that of a beginner. *Break Down* was the artistic process made visible.

The work of Marilynne Robinson is the artistic process made transcendent, redemptive, inward. In her novels and essays she defends the truth and the irreducible reality of the inner life of our minds against the reductive spirit of the age. I am pretty sure I have met three geniuses in my life: Samuel Beckett, John Ashbery and Marilynne Robinson.

When I last saw her, she was, unexpectedly, navigating her way back to her London hotel using the location system on her iPhone. She was in town for the Orange Prize for Fiction, which, a couple of days later, she would win for her novel *Home*. I had previously spent some time with her at her home in Iowa City and acquired the very clear conviction that technology had, for the most part, passed her by.

The house was neat and wooden and on the green fringes of the university district. We sat for two or three hours on the large verandah. Rain fell intermittently, birds sang unusually loudly and a distant train whistle blew, one of the great soulful sounds of America. Robinson was sitting on a 'porch swing', a hanging sofa, and she swung continuously throughout our conversation, an odd, gentle echo of Bill Gates rocking in his chair.

She laughed a lot and her whole body shook when she did. She wore a silver bracelet and an enormous watch. She has thick, grey-brown hair, a round, kind face and very watchful eyes. There is something slightly nervous, even girly, about her. She says 'you know' in the middle of almost every sentence to mean she does but you may not. Her formidable scholarship means this often happened. She is passionate about Obama, John Calvin, the history of Iowa and the Congregationalist church in Iowa City. She likes Willie Nelson and Emmylou Harris. She teaches at the Iowa Writers' Workshop and loves her students, though she seldom reads their work after they leave. Later in a bar, over a couple of beers, she became intense and derisive about the idiocies of certain scientists.

In 1980 she published a novel, *Housekeeping*. It was critically praised, won awards and was filmed by Bill Forsyth. A substantial new novelist had arrived. But she didn't publish another novel for twenty-four years. Instead, she wrote a non-fiction book on British nuclear policy called *Mother Country* and issued a collection of essays called *The Death of Adam*, Then, finally, in 2004 she published another novel, *Gilead*: more critical acclaim, more awards. And then, only four years later, she wrote *Home*. It is the story of *Gilead* seen though different eyes. Later she was to publish a series of lectures under the title 'Absence of Mind'. They were an attack on the denial of the importance of our inner experience in so much contemporary thought, primarily among scientists.

She was born and brought up in Idaho, the setting of *Housekeeping*. The beauty of the place, she says, has always been valuable to her. It was a middle-class home and she says, tentatively, she was happy, but then she seems to withdraw from this claim. 'I think I've never been precisely suited to life in the world. It took me a while to get the composure and wherewithal to figure out how I wanted to be situated in the world and I think my childhood may have contained an element of bewilderment.'

It wasn't an especially religious upbringing.

'I never doubted the religious seriousness of my family but they were never particularly explicit about it . . . To a very considerable degree I'm a religious autodidact, but I don't know if my mother would want me to say that.'

At college – Brown in Rhode Island – she was introduced to theology. She started writing in high school and began a novel in college, but the moment she graduated she loathed it. Her Ph.D. was on Shakespeare. She married in 1967, had two sons and subsequently divorced.

She started *Housekeeping* when the family was living in France in the late 1970s. 'I went into a bedroom and closed the shutters. I had a little tiny lamp and my spiral notebook. It was like a sensory deprivation chamber ... It was an uncanny experience, I found I could retrieve things from memory with a vividness I had never anticipated. It's always been my habit to more or less trust my memory to make my choices for me.'

The language of the book has an unfamiliar solidity and glow. Like all her novels, *Housekeeping* doesn't seem to have been written at all; the sentences seem to have been there for ever, waiting to be discovered.

This effect goes to the heart of her greatness as a writer. She rejects the realist conventions of the novel. 'I think the assumptions of realism as it has been practised are simply wrong. People bring a great deal of memory and also a sense of present experience to everything that they do. If you see someone doing a simple action like hanging sheets on a line, there is absolutely no reason in that person's perception that there is anything simple about it at all. I have all the respect in the world for reality but I think the general assumptions about it are wrong.'

She thinks in metaphors because everything is a metaphor. This is her faith – the world not as a factual cul-de-sac but as an unfolding revelation, of constant newness. In her essays, this extends to inspiring attacks on the reduced view of humanity offered by contemporary science – in particular, the culturally illiterate view of religious imagery.

'For heaven's sake, the idea that the dome of the sky is the skull of a murdered god. What is being described there? A very great deal. The idea that that is the kind of statement that could be displaced by something about gravity or the atmosphere – that's a bizarre assumption to make ... At a certain point in cultural history there appeared this idea that people are biological automatons and everything to do with perception and emotion and birth and death is some sort of epiphenomenal thing that should be excluded from the definition of the real. This, to me, is very bizarre.'

She arrived at the Iowa Workshop, already established as one of the most distinguished writers' courses in the world, in 1989. She has become its figurehead. Students pack her seminars. Strangely, she dislikes being called a writer. 'I think "writer" is a toxic word. I'm a writer when I'm writing something. The rest of the time I like to put that word aside.'

In 1998 she published *The Death of Adam*, a monumental series of essays on faith, science and, well, everything. It was the product of her long immersion in poetry, science and history. And then, in 2004, came *Gilead*.

An old man preparing for his death, John Ames, an Iowa minister, is writing a long letter to his young son. His memories take in the history of settlement, slavery, the deep conflicts that lie beneath the now peaceful soil of her adopted state. Ames is ready for the afterlife, but he is so in love with this world, he cannot bear to leave it. He is a rarity in literature – an utterly convincing great and good man. *Home* is the same story told in a different way. She found she couldn't leave the *Gilead* character of Jack alone. Jack is the son of another minister, a good man but a bad sinner.

'I'm very fond of him ... I didn't want to make Jack a good man in a conventional sense, I wanted to make him a person of value in terms of the whole complexity of his life.'

She wanted to see him redeemed.

It grew cold on the verandah and she wanted to take me to see her church and her minister, Revd Bill Lovin of the Congregational United Church of Christ. She slumped contentedly into a big chair in his office – 'I could sit here forever' – while smart, genial Revd Lovin and I discussed American religion.

Her language, her sense of the inner life, of the preciousness of personhood and the possibility of redemption, of making anew, of starting again, are all very American. Fitzgerald would have recognised a soulmate in Robinson. And so, in a very different way, would Saul Bellow. In his late novella *The Actual* (1997) Harry and Amy were almost lovers once but she went off and married the hopeless Jay, a bad lawyer and a shallow man. Years later Harry and Amy meet again. This is real love, the spiritual key for which Harry has been searching.

'I stood back from myself and looked into Amy's face. No one else on all this earth had such features. This was the most amazing thing in the life of the world.'

Harry looks into the mystery of personhood and realises it is everything.

But the personality of art is not just an American mystery. Mozart also wondered about this after he had described his com-position process in that letter quoted by Hadamar. If the work seemed to exist before it came into his mind and if he was just the means of its appearance in the world, how could it be a piece by Mozart and nobody else? How could it have any special personality at all?

'Now, how does it happen,' he wrote, 'that, while I am at work, my compositions assume the form or the style which characterize Mozart and are not like anybody else's? Just as it happens that my nose is big and hooked, Mozart's nose and not another man's. I do not aim at originality and I should be much at a loss to describe my style. It is quite natural that people who really have something particular about them should be different from each other on the outside as well as on the inside.'

It is not a very clear answer, but, then, 'How does it happen?' is not a very clear question. Art is eternally new, eternally arriving and it is both universal and highly personal. It escapes all our secular, scientific categories. It is as complex as we are.

On the verandah, the sound, her swinging and the damp, dim light made me feel that we have been sitting there for ever. In a way, in Marilynne Robinson's faith, we had, for this was heaven, an always renewable reality.

'Even in Revelations,' she said, 'what is promised is a new heaven and a new earth, which sort of suggests that what we have here ... If we were going to fix our imagination on what awaits, the best evidence is what we have here and, granting the painful circumstances under which many people live, if you think of this reality renewed it would be a very splendid thing, I would not ask for more.

'It's a vastly more moving idea to me than the idea of pearly gates. I mean this' – she sweeps her hand in a gesture that takes in the house and the deep green, soaking garden – 'would be heaven enough for me.'

Robinson's new heaven and new earth are Fitzgerald's 'something commensurate to his capacity for wonder', they are the world remade in an instant. They are what it feels like to be alive for at least as long as we choose not to be simplified into machines.

13

THE AGE OF COMPLEXITY

Against the seductive, simplifying forces of the new machine age and reductive accounts of the human mind stand not only creativity and art, but also new kinds of science and new ways of seeing. The beginner's mind of the artist is matched by the beginner's mind of the scientist who sees the world anew, accepting its complexity without wishing to change or improve it. A new scientific vision is the subject of this chapter.

In this book I have told stories of the power of crude simplifications, simple solutions to complex problems. There were the naive, early forecasts of artificial intelligence and robotics research. There was the surgery that took away Henry Molaison's memory. There are the call trees that make us miserable or the banks clinging to their mystical belief that their equations control risk. There is the celebrity culture that makes robots out of humans. There are the superstitions attached to neuroscience and to the gadgets that daily demand our attention, our money and our identities. There are the rabid gamers and the worship of information for its own sake. There are the children with fake friends and the parents absorbed in their BlackBerries. There are the fantasists dreaming of immortality via the upload and there is the cult of the future and of technological determinism embodied in the myth of the Singularity.

But there is a different type of story that begins with trillions of strange animals that expose the folly of thinking we know more than we do. These are the extremophiles, one of the most bizarre and heartening revelations of modern science. They were first discovered in the 1970s living in the hot springs of Yellowstone National Park and they have since been found in just about every improbable corner of the planet.

Extremophiles are organisms – mainly microbes – that live in conditions that would previously have been considered too extreme to support life. We have found acidophiles that live in intensely acid environments, thermophiles that thrive on intense heat, psychrophiles that are equally attached to deep cold, xerophiles that can survive with almost no water, barophiles that can live in pressures of four hundred atmospheres and anaerobes that require no oxygen.

But the extremophile that people seem to find the most loveable is *Deinococcus radiodurans*. It is certainly astounding evidence of the sheer tenacity of life. This bacterium is so resistant to radioactivity that it can live in the cores of nuclear reactors. DR achieves this by a seemingly miraculous process whereby it continually repairs radiation damage to its own DNA. Why this mechanism should have evolved in the absence of such extreme levels of radiation on earth is a mystery.

That we should have taken so long to discover extremophiles is strange. They are not rare, they are utterly ubiquitous. In fact, the total weight of extremophiles probably exceeds the total weight of all other living things. And yet we were blind to them – we thought we knew so much until, suddenly, we discovered we knew next to nothing. That fact alone is a warning to all believers in simple interventions in nature.

The most startling – in terms of their sheer scale – extremophiles have only recently been discovered. These are microbes that live beneath the seabed. Living cells have been found 1,600 metres beneath the ocean floor. A place we previously thought was dead has turned out to be the largest ecosystem on the planet.

John Delaney is professor of oceanography at the University of Washington. A bearded, impressive figure with a touch of the nineteenth-century preacher about him, he talks about the oceans as, among other things, 'the last frontier on earth' – a very American concept – and 'the largest, most complex biome on earth' – a biome is a large, living community of distinctive organisms.

He works on the Ocean Observatories Initiative. There is a sort of double take on the word 'observatories' when used in this context, as there is with the Brain Observatory in San Diego where Henry Molaison's brain was sent to be cut into slices. Observatories are usually astronomical, designed to reveal the unfamiliar and, to unaided eyes, invisible. But the seas are all around us in plain sight; they are utterly familiar. Why do we need observatories?

Delaney answers that question by quoting Marcel Proust – 'The real voyage of discovery consists not in seeking new landscapes, but in having new eyes.' The new eyes – another way of describing the beginner's mind – in this case consist of a menagerie of censors, robots and internet links with which we can explore, potentially, the entire abyssal benthos – the beautiful technical term for the deep ocean. 'Abyssal' refers to the abyssopelagic layer which ranges in depths from 4,000 to 6,000 metres and 'benthos' is the term for life on or near the seabed.

A sci-fi landscape apparently showing exploration of an alien planet appears in Delaney's presentations. Elevators rise and fall from the seabed on cables that moor flying saucer-like research stations. Robot submarines cruise over the seabed which is littered with microbial incubators and pressure sensors.

'We will,' says Delaney, 'be present throughout the volume of the ocean in real time.'

Through ecogenomics, the DNA of organisms on the sea floor will be read and the information immediately passed back to laboratories on land. Robots will fly to erupting undersea volcanoes and collect the microbes that emerge. A float will be sent to the sea surface, picked up by a robot plane and the microbes will be in the lab no more than twenty-four hours after they emerge.

There are two big practical reasons for doing this. Earth systems are driven by two forces – the sun and the internal energy locked up beneath the crust. The seabed is the place where we can see the interactions between the two. Climate science could be transformed by what we discover – Delaney describes the oceans as 'the flywheel of planetary climate'. Secondly – another Delaney-ism – the deep ocean may well be the next rainforest for the pharmaceutical industry, yielding new compounds to fight disease.

But there are also more poetic reasons, not least the desire to know more about something as important and yet as unexplored as the sea.

'This is the system,' says Delaney, 'this is the crucible out of which life in the planet came.'

He is very fond of poetry. In a lecture he tells the story of Matsuo Basho, the great master of the haiku, and quotes at length from T. S. Eliot's *Four Quartets*. On his university website he has eleven poems including Matthew Arnold's 'Dover Beach', Pablo Neruda's 'It Is Born' and Robert Frost's 'Once by the Pacific'. He also quotes Basho's haiku 'Turbulent', which, significantly and with breathtaking beauty, links the unexplored ocean with unex-plored space:

Turbulent the sea
Stretching across to Sado
The Milky Way

Basho once said, 'If you would know the pine tree go to the pine tree'. This precisely echoes an answer the Russian writer Anton Chekhov gave to his wife's question, 'What is the meaning of life?' The dying Chekhov said this was like asking what a carrot is – 'A carrot,' he explained, 'is a carrot and nothing more is known.'

In our scientific vanity we think we can discover the pine tree and the carrot by taking them apart, breaking them down into their constituent parts. But, at the end of the process, they remain untouched – a pine tree is a pine tree and a carrot a carrot, complex and irreducible, like life. There is no point in trying to discover the pine tree or the carrot in sub-atomic particles or in DNA; you can only know either by seeing them as what they actually are in the world.

But Delaney's primary impulse seems to be that of the explorer – to travel to distant places just because they are there. Or, rather, to find ourselves. He quotes from *Four Quartets*.

We shall not cease from exploration
And the end of all our exploring
Will be to arrive where we started
And know the place for the first time.

Seeing complexity is the first step to accepting it, to knowing it for the first time. Observing the oceans is one such step, observing the whole planet was another. Seeing earth from space has been one of the historic privileges of my generation. The first photograph of the whole planet from a spaceship appeared in 1966. NASA's unmanned Lunar Orbiter 1 was about to pass behind the moon when, in an unplanned manoeuvre, mission controllers pointed its camera away from the moon and towards the earth. Then came 'Earthrise'. It was taken on Christmas Eve 1968 by a human being rather than a machine – Bill Anders, lunar module pilot on Apollo 8.

The picture is one of the most important images of our time, not because it was the first but because, aesthetically, it was the best. Like Delaney observing the deep ocean and seeing its complexity, Anders dramatised something we already knew – the complexity of the living planet in an abyss of rocks and emptiness – and made us see it for the first time.

'Earthrise', taken with a Hasselblad camera with 70mm film,

mostly consists of the absolute black of space. The grey lunar surface and the slanting horizon line occupy the lower fifth of the frame. Just above the horizon, about two-fifths of the way up the frame and offset slightly to the right, is the earth. It is a hemisphere, the lower edge of which fades into the surrounding blackness. The surface is blue and covered with coils, streaks and splashes of white cloud. The slant of the curve of the hemisphere's lower edge echoes the slant of the horizon line, but not quite: the angle is slightly more acute. The earth, the picture made clear, is different.

The true wonder of space travel, in our time at least, is aesthetic. Anders' photograph gave us the beginner's mind. We genuinely were seeing something for the first time and we were seeing earth whole. It was impossible after that picture not to see the planet as a complex and miraculously emergent system.

The picture also reminded us of our dependence on that system. This is not just a matter of seeing the earth as a small planet in a big universe; it is a matter of realising that this is our home, our point of origin, and we have no other. Leaving home is hard. Space travel is enormously difficult because all journeys are long and because humans are not designed to live in a zero-gravity, highly radiated vacuum. The sheer engineering problems of living in space are such that it is doubtful – in spite of the hopes of the cyborg dreamers – whether it can ever be achieved.

The resulting costs and the uncertain benefits of space have meant that the government space programmes that generated such excitement in the 1950s and 1960s have lost almost all their imaginative potency. Now, after decades of frustration, they are being replaced by private ventures.

Burt Rutan, born in Oregon in 1943, is a very independent aerospace engineer with an uncommonly good eye. The specialised aircraft he designs look like no others and they are extremely beautiful. In 1982 he founded Scaled Composites, a company that became one of the leading design and experimental aerospace companies in the world. It is based at Mojave Airport, a civilian testing site surrounded by very secret military sites in the Mojave Desert in Southern California, and it was there that I met him in 2005.

'It's not rocket science' is often said to suggest that something is very simple. The point being that rocket science is seen as very difficult. In fact, it isn't.

'A rocket,' said Rutan, 'is a bomb with a hole in one end.'

What is difficult is the paraphernalia – life-support systems, aerodynamics, control mechanisms – associated with rocketry designed to carry humans into space. But Rutan seemed to have solved most of these problems remarkably cheaply.

Richard Branson's Virgin had invested in Rutan's company after his SpaceShipOne won the Ansari X Prize, a $10-million-prize for the first private sector team to put a ship in space twice in fourteen days. SpaceShipOne was a rocket-powered plane launched in the air from White Knight, a craft so strange and elegant that I had to ask Rutan why it was that shape.

'Why not?' he replied.

Rutan won the prize on a budget of $26 million; at the time (2005) NASA's annual budget was $16.2 billion. He was financed by Paul Allen, the co-founder, with Bill Gates, of Microsoft. Rutan was, unsurprisingly, scornful of NASA, especially of the Space Shuttle, the ugly vehicle that replaced the glorious Apollo rockets that had taken men to the moon. 'They should have gone into a smoke-filled room before the Shuttle flew and admitted to themselves that they'd fucked up.'

NASA had told the US government that the Shuttle would make it ten times cheaper to put payloads into space than the Apollo's Saturn V rocket; in fact, it was ten times more expensive. It had also been estimated they would lose Shuttles at the rate of one every 100,000 missions; in fact, they lost two in a little more than one hundred missions. Rutan looks on the era of paralysis caused by the Space Shuttle with dismay and contempt and he believes that, now the private sector has begun to break the state monopoly, the real golden era of space is about to begin. 'In 1908 Wilbur Wright flew his aeroplane in Paris and the whole world started to look at it differently. They thought: if this guy who owns a bicycle shop in Dayton can do it, then so can we. Within four years there were hundreds of different types of planes in thirty-nine different countries. There were two factories in Paris that built five hundred planes! All from nothing in four years!'

There is, however, a fundamental difference between aeroplanes and spaceships and it is not just cost. Aeroplanes take you from A to B and back again; spaceships don't take you anywhere except space, which is why the cost of space travel has become ever harder to justify.

But the possibility still grips the imagination. Aesthetic wonder has been enough to generate disappointment and then impatience with the failures of state-sponsored rocketry. It is almost forty years since Apollo 17 blasted off from the lunar surface and, since then, no human has gone beyond earth orbit. While every other technology has raced ahead since the 1960s, space technology seems to have stalled. So it is perhaps not surprising that some of the princes of Silicon Valley have decided to apply their skills to space.

Elon Musk, the model for the wayward capitalist Tony Stark in the *Iron Man* movies, was born in 1971 in South Africa and moved to Canada in his teenage years and then to the US, arriving at Stanford University in Silicon Valley in the 1990s, just at the start of the first internet boom. He lasted two days at Stanford and left to form his first company, Zip2 Corporation, which provided online city guides. In 1999 the company was bought by Compaq and Musk walked away with $22 million. Then, via two other companies, he acquired and then transformed PayPal, which organises payments through the internet. PayPal grew rapidly thanks to its link with eBay and, in 2002, Musk sold it to the online auction company for $1.5 billion of which his share was $165 million.

Also in 2002, he founded SpaceX with the intention of sending an experimental greenhouse to Mars in which plants would be grown in a special nutrient gel. He reckoned he could do this for $20 million but then discovered that using an existing rocket – Boeing's Delta – would add a further $50 million to the cost. So he decided to build his own.

His rockets – called Falcons after Han Solo's Millennium Falcon in the *Star Wars* films – are cheap and they work. Falcon 1 put a satellite into orbit. There is also Falcon 9 which can be upgraded to be a full-scale heavy-lift rocket and will be used to deliver cargo and, potentially astronauts, to the International Space Station (ISS) and, thereafter, to Mars. He even has a capsule, the SpaceX Dragon module, which would carry people. The company has won a $1.6 billion NASA contract for twelve Falcon 9 flights to the ISS. But Musk's Mars ambitions are still in place.

'One of the long-term goals of SpaceX,' Musk has said, 'is, ultimately, to get the price of transporting people and product to Mars to be low enough and with a high enough reliability that if somebody wanted to sell all their belongings and move to a new planet and forge a new civilisation they could do so.'

The project would take twenty years, far quicker than anything now likely to come from state space plans, but, to Musk, it is a 'semi-infinity'.

Rutan, Branson and Musk have acquired many other competitors. From computer culture, there is Jeff Bezos, founder of Amazon.com, who set up Blue Origin in 2000, which is building New Shepard, a rocket designed to take space tourists up to 400,000 feet – almost eighty miles. And there is John Carmack, the lead programmer of computer games like *Quake* and *Doom*, who has established Armadillo Aerospace, which is also developing a space tourism vehicle as well as a lunar lander. Then there is Jeff Greason of XCOR Aerospace and, in Britain, Steve Bennett with Starchaser, another suborbital tourism programme.

The list keeps lengthening. In early 2011 it was announced

that twenty-nine teams had signed up for the $30 million Google Lunar X Prize. The first prize of $20 million goes to the team that, by 2015, lands a robot on the moon which travels at least 500 metres and sends back high-definition images.

It is no accident that Musk, Bezos and Carmack made their fortunes from the internet and computer games and that Google is backing the Lunar X Prize. Those new technologies are often seen as new frontiers, but the word is being used metaphorically. Space is not a metaphorical frontier; it's the real thing and the wonder it inspires is – or should be – humble rather than imperious. It is an exploration and discovery of nature. The simple rocket helps us to see the complexity involved in our survival on this planet. Of course, exploitation will come later, but, for the moment, the goal is discovery.

But rockets and ocean observatories omit the human. Seeing people has been the job of anthropologists, a discipline that has now been invaded by the concept of complexity. The insights produced by this invasion should be celebrated in the near future by the award of the title World Heritage Site to the rice plantations of Bali which have existed, more or less unchanged, for at least a thousand years

We know the time frame because we know that, in the year 884 in the Saka calendar or AD 962 in the Christian, King Chandrabhaya Singhawarmadewa of Bali ordered the restoration of the Pura Tirtha Empul, the Temple of Holy Water. The order was inscribed in stone. It survives as the earliest written record of this temple. There are many water temples in Bali. Each is a gift from the goddess Dewi Danu. They form part of the *subak* system for managing the rice terraces that rise up the slopes of the island's volcano, Mount Agung. *Subak* is a word that first appeared in the eleventh century. It refers to the system for sharing water among the rice farmers on the slopes of the volcano. The farmers own their land, but the *subaks* are common property. Dewi Dani entrusts the management of her water to the farmers on condition they honour her in special rituals conducted in the water temples.

In the mid-1970s the Green Revolution came to Bali. The great agronomist Norman Borlaug was the prophet of this revolution. It has been estimated that Borlaug saved a billion people from starvation by promoting the planting of high-yield crops and new agricultural techniques to feed a soaring world population. His revolution swept the world.

Steve Lansing, professor of anthropology at the University of Arizona at Tucson, first went to Bali in 1971 as an undergraduate, returning in 1974 as a graduate student. He had been going to study physics but switched to anthropology. 'I was interested,' he says, 'in the temples and the culture. The more I learned the language, the more I realised they were having a great struggle with this Green Revolution problem. They had been given a programme called Massive Guidance. The idea was to set up an infrastructure so that the farmers would be given what they needed. They called them technology packages – rice, fertiliser, pesticides – on credit each planting season. And then, at the end of the planting season, they would buy the rice from them. So they would get money into the hands of the farmers and increase production. That was the Green Revolution programme every-where and it worked.'

But then he adds: 'It just happened to neglect what was going on on the ground and the way the Balinese had organised things for the previous thousand years.'

The mystery about Bali was that it seemed to defy what many took to be an iron law of politics and sociology. In 1968 an ecologist called Garrett Hardin laid down this law in an article in the journal *Science* entitled 'The Tragedy of the Commons'. The essay, like the Green Revolution, sprang from the anxieties of the age, primarily the fear of over-population. It is one of the defining documents of our time. This is not just because of its argument but also because it captures the spirit of our age – globally anxious, fearful of the future, stricken by the revelation of the perils of modernity and gloomy about the limits of rationality.

The tragedy of the commons happens because, when land or any other resource, is open to all, it is in the interests of each individual user to exploit it as ruthlessly as possible. In time, this will destroy or deplete the common resource.

For Hardin, this meant that there needed to be control. In the context of an over-crowded world – Hardin's primary concern – fertility would have to be controlled by law. The tragedy of the commons should have meant that Bali's *subak* system could not work because nobody was taking the sort of rational decisions Hardin advocated. But it did.

Struck by the Balinese farmers' unease with the advice of the Green Revolutionaries, Steve Lansing looked closer at the *subak* system. It was, he realised, a very subtle arrangement involving not just agriculture but the social system and the local religion. This religion – part Buddhist, part Hindu, part Polynesian – was centred on Tri Hita Karana, or three causes of goodness. The three causes were harmonious relationships with the human world, the spirit world and nature.

For Lansing it could not just be a process of asking the farmers how it worked. They did not know; they just knew it did work. 'Everybody knew what was on the ground in front of them but they hadn't really thought about the larger structural issues. Why would they, right? As long as it was working for them, it was nobody's job to understand how the traditional system worked.'

Eventually, in the early 1980s, Lansing modelled rice growing in Bali on a computer. He discovered that the farmers' system was as good as it could get. They had arrived at the optimum use of the land. There was nothing the Green Revolutionaries could teach them. I asked Lansing how the farmers reacted when he told them. 'With delight! It shows that they know what they are doing, it's a kind of validation. They were fascinated. It provides a way of looking at what they are doing already which they wouldn't otherwise see.'

The computer model revealed something even more extraordinary. It showed they could have arrived at this system not after a thousand years but after ten years. Put all the elements in place and the model raced to the Bali outcome. Nobody thought of this in advance, it just happened. Enduring stability was an emergent property of the system. The tragedy of the commons had been averted not by laws but by a shared world view.

A shared world view is something that we, in the contemporary West, lack. Western humanism has become radical individualism in which every person's world view is given equal status. But we have an option that springs from an awareness of the complex system by which we live. True environmentalism supersedes humanist values because it accepts our dependence on the entire living system whose values are far more important than our own. One man, a friend of mine, sees this more clearly than anybody.

On 26 July 2009 James Lovelock was ninety. A party was held in the orangery of Blenheim Palace in Woodstock, near Oxford. It was supposed to be a surprise for Jim, but, in the event, it wasn't. He is a penetratingly observant man and his softly spoken wife and muse, Sandy, could not conceal the party planning from his analytical gaze. The great and the good, grand figures from science, politics and the arts, turned up to honour him. He was a man vindicated.

He is also a man who never ceases to amaze. In his speech at the party, he did not dwell, as most people at his age would, on the past. Instead, he looked to his next big adventure. Now ninety-two, he has a seat booked on one of Branson and Rutan's first space flights on SpaceShipTwo.

I first met him some years before at his home on the Devon/Cornwall border. Buried deep in the countryside, it is very hard to find. This is a farmed landscape rather than a wilderness but the search for Jim down narrow lanes still feels like a descent into wild England. His home is a workshop, laboratory and experimental station, a sign warning of radioactivity is on his front gate. He is not only a scientist, he is an engineer, making his own devices. This talent made him the equivalent of Q, the man behind James Bond's gadgets in the books and movies. Jim is a patriot; he has worked for the intelligence services.

But one thing he does not do is try and control nature: it runs wild and free over his land. He did once attempt systematic tree planting, but then realised that nature knew better and he now leaves the land to sort out its own planting scheme. Her gardening schemes are, for him, an endless source of delight.

At our first meeting, he showed me a small tube and just said, 'That's it!'. The tube is called an electron capture detector (ECD); Jim invented and built it in 1957. It is one of the most important gadgets of our time. The fact that he did it all himself is important. He is an entirely independent scientist, unconnected to any institution. This gives him a certain defiance, a useful quality since he has had to be defiant to defend his single biggest idea against the most withering attacks. An authorised biography of him is called *He Knew He Was Right*. He did and he was.

All the ECD does is detect traces of chemical in the air – or any gas – but it does so with staggering precision. Lovelock once told me that if a blanket suffused with a chemical was waved about in Tokyo, he could detect traces in the Cornish air two weeks later.

In the 1960s, the ECD revealed for the first time the threat humans posed to the earth. It showed that pesticide residues were pervasive throughout the atmosphere and that CFCs – chemicals once commonly used in fridges and aerosols – had polluted the planet. Along with Rachel Carson's book *Silent Spring*, published in 1962, Lovelock's little tube launched the contemporary environmental movement.

Carson observed birds, Lovelock analysed the air. The pesticide DDT penetrated the food chain and killed birds; CFC molecules floated high in the atmosphere and destroyed the ozone layer which protects life from the effects of high-frequency ultraviolet light from the sun. The conclusion leapt out of the basic science: human activity could fundamentally change the life-support systems of the entire planet. Our simple fixes were capable of destabilising or destroying the complex system that gives us life.

DDT and CFCs were simple technical solutions intended to make human life more efficient and liveable and that is indeed what they did. But they did so at a cost, one that was intrinsically unknowable as it is impossible to predict the effects of continued destruction of the ozone layer and of the destruction of a large part of the biosphere. The revelation was that simple solutions have unpredictable effects on complex systems and nothing, not even the human brain, can compare in complexity to the living system of the earth. The insight of the environmental movement is the great insight of our time – we meddle with complex systems at our peril.

Lovelock once told me that it takes forty years for an idea to be understood and accepted. He was a decade too optimistic. The full significance of the importance – and, indeed, sanctity – of complex systems has only now started to be fully understood. The leading scientists in the world are now beginning to treat nature not as something to be fixed but as something that should be understood and sustained.

In the 1960s Lovelock was asked by NASA to suggest ways of detecting life on Mars. It was generally thought that we would simply fly to Mars, scoop up some soil and examine it for signs of life. After all, earth soil is teeming with life. The underlying assumption is that life on Mars would be like life on earth which is, on closer examination, a preposterous idea. So what if, by scooping up soil, we were looking for the wrong thing in the wrong place?

'Ordinarily,' Lovelock wrote in a paper on his work, 'one does not look for fish in a desert, nor for cacti on an ice cap.'

His point was that you have to know what you are looking for in the place you are looking. His answer to NASA's question was: we should look for order and non-equilibrium. Life, he argued, did not passively evolve in response to an environment; it changed the environment – as *Silent Spring* and the ECD had shown. The first thing an alien studying earth would notice is that our atmosphere is unstable and improbably ordered. This is because of life, which turns the entire planet into a dynamic, complex system. If there were no life it would have settled into thermodynamic and chemical equilibrium, the atmosphere would be stable and disordered. The earth does not just contain life, it is alive.

And he really did mean alive. Following his work for NASA, he formulated the Gaia Hypothesis which argued that the earth was an integrated, complex, interactive system that behaved like a single organism. Just as the human body has multiple self-regulating systems, so the living earth cares for itself through planetary scale interactions. These could only be properly understood if we considered it as a single system, like an organism.

Gaia was the Greek goddess of earth and the name was suggested to him by his friend the novelist William Golding. The name was seized upon by critics of the theory who tend to dismiss the hypothesis as quasi-religious, New Age nonsense. The critics failed and Lovelock's idea – often uncredited – is now at the heart of all earth systems science. It is also at the heart of a new conception of our place not just in space, but in time.

The Long Now Foundation (LNF) was born in 1996 or, for reasons I shall come to, 01996. It was set up by, among others, the musician Brian Eno, biologist and creator of *The Whole Earth Catalog* Stewart Brand, and Danny Hillis, computer engineer and inventor. The foundation's function is to promote long-term thinking. The Long Now – a name which came from a remark of Eno's – stands in opposition to the short now, the refusal to consider the future as anything more than a few years – or a few minutes – ahead.

'The future,' said Danny Hillis, 'has been shrinking by one year per year for my entire life. I think it is time for us to start a long-term project that gets people thinking past the mental barrier of an ever-shortening future.'

The foundation, based in San Francisco, runs a number of projects. Long Bets encourages people to make long-term wagers. Participants put up propositions – like 'By 2060 the population of humans on earth will be less than it is now' – and then people are encouraged to bet for or against them. The Long Server is an attempt to create a complete back-up of digital information. The Rosetta Project is intended to preserve all languages that might go extinct within the next hundred years.

But their most thrilling and strange project is The Clock of the Long Now. I go searching for the prototype of this clock in the Science Museum in London. It is hard to find. The man at the information desk at the Science Museum sends me to the time gallery on the first floor. But it is not there and the time lady sends me back down to the first floor where there is a gallery called The Making of the Modern World. Even here, among these solemn glass cases full of significant machines, it is elusive.

Finally, I stumble on an unusually large vitrine with a faintly neglected air. One of the four uplights inside has failed and clumps and tangles of dust lie on the upper glass surface. The clock looks, amid all the flashing lights and fun of the museum, forgotten. Parties of chattering schoolchildren pass without a glance; they are on their way to more exciting, more obvious exhibits.

No wonder it doesn't make them pause – the object inside is enigmatic, opaque. It is large – eight or nine feet high – and made of heavy, industrial age materials. In form, it seems to echo both the heavy fragment of Charles Babbage's Difference Engine 1, which lies, unfinished forever, in its own vitrine a few yards away, and his Difference Engine 2 upstairs in the mathematics gallery. Also, unlike the interactive electronic gizmos scattered throughout

the museum, these mechanical computers look entirely inaccessible, unresponsive.

The machine is not old; it was completed in 1999. It is, however,
designed to become very old, as old as history. Humans settled
and became farmers – and historians – 10,000 years ago. This
object is the first prototype of a machine that will survive, intact
and working, for the next 10,000 years.

The rare visitor who stops to look is confronted by a face
consisting of six dials representing the year, the century, horizons,
sun position, lunar phase and the night sky. A torsional pendulum
with a one-minute period keeps the time. The two columns on
either side are helical weight drives which power the mechanism.

The final version will be monumental, vast – sixty feet tall.
Power may be constantly provided by the footfalls of visitors
and changes in temperature or the clock may be wound in the
traditional manner. Its accuracy will be constantly checked by a
solar synchroniser that, when the skies are cloudless, consults the
position of the sun at noon to check the correct time.

The year now, according to this clock, is not 2011 but 02011,
the initial zero indicating how much time there is left to run. This
clock is an optimist, it expresses faith and hope. Our species, says
the clock, will survive to locate itself in time with a five-figure
year.

Potential sites for this clock are being investigated near Van
Horn in west Texas and Mount Washington in eastern Nevada.
The Texas site, which is on land owned by Jeff Bezos, is intended
as a trial run for the full-size clock. Nevada will be the final, public
version.

The Clock of the Long Now will be a perspective device. Like
a telescope or a microscope, it is intended to make us see things
differently. The 'now' of the clock is not that of this moment, nor
even that of a lifetime; it is that of the entire human moment. The
clock will be, says Stewart Brand, 'a mechanism of myth'.

The Victorians were horrified by the geologists' discovery of deep time – the true age of the earth – and by deep space – the unimaginable scale of the universe. They felt lost. We are now trying to find ourselves as enduring and significant creatures in spite of or even because of the depths of space and time.

Short-term thinking, as Hillis implies, threatens our survival. We invent nuclear weapons and burn carbon to solve short-term problems, only later discovering these things can acquire a destructive logic of their own.

This is where the mind of Jim Lovelock intersects with the minds of the Long Now. He fears we may, as a result of environmental catastrophe, lose everything. At the end of his book *The Revenge of Gaia*, he argues that we live in adversarial rather than thoughtful times and that science has become too specialised, too arcane. We lack clear statements of the truth of our condition at a time when, thanks to global warming, we may be facing the complete collapse of our civilisations and the reduction of our species to a few warring tribes in the areas of the planet where we can still survive. He suggests it is time to assemble a book – 'a guidebook for our survivors to help them rebuild civilization without repeating too many of our mistakes'.

Current books, he says, take scientific knowledge for granted and, in the shops, equal space is given to astrology, creationism and homeopathy. 'Imagine the survivors of a failed civilization. Imagine them trying to cope with a cholera epidemic using knowledge gathered from a tattered book on alternative medicine.'

Jim says we need a beautifully written book of knowledge – 'a manual for living well and for survival . . . a primer of philosophy and science'. It would tell its readers how to light fires as well as the earth's place in the solar system. The book would explain infectious diseases and the circulation of the blood. It would be about art and civilisation. It would be a new bible, distributed, before the catastrophe, to schools around the world. We have come so far in this Long Now; it would be foolish to throw away all we know.

Men like Brand, Hillis, Lovelock, Delaney and Lansing, and like the new space explorers, are paradoxical figures. They are hard scientists or thoroughgoing materialists who are nevertheless filled with a sense of the poetry in what they do. This is perhaps because of the way they see, like T. S. Eliot, true simplicity lying only on the far side of complexity.

For Jim, Gaia is the ultimate form of complex simplicity. He has recently changed his mind about his theory. It is usually assumed that Gaia is a metaphor, that the earth is not really a single organism, not least because it cannot reproduce. This is what Jim used to think, but, latterly, he has decided Gaia is not a metaphor, that the earth really is a single living system. The reproduction argument is not a problem, he says, because he doesn't see the need for a four-billion-year-old organism to reproduce. The goddess is real, we are all part of her, the Long Now is not 10,000 years but four billion, the moment of the living earth.

Jim wants to look into the face of this ancient goddess, but, agonisingly, the Virgin Galactic flights have suffered delays. He is still awaiting the call as I write this, still, at ninety-two, determined to go. He has just emailed me.

'All I want to see,' he wrote, 'is Gaia.'

Gaia is the supreme complexity. She tells us we must be humble and careful. The old age of the short-term, simple solution is dying to be replaced with something much finer – the Long Now of the complex system.

THE PARIS HILTON PROBLEM

But we remain human, all too human.

In 2010 a leader in *The Economist* warned of the dangers of cyberwar. 'The threat is complex, multifaceted and potentially very dangerous. Modern societies are ever more reliant on computer systems linked to the internet, giving enemies more avenues of attack. If power stations, refineries, banks and air-traffic-control systems were brought down, people would lose their lives.'

Cyberweapons, warned the leader writer, were difficult to monitor and, unlike nuclear arsenals, impossible to verify. They are also, in current strategic jargon, 'asymmetric'. Lone computer hackers can take on superpowers. 'But the world needs cyberarms-control, as well as cyber-deterrence. America has until recently resisted weapons treaties for cyberspace for fear that they could lead to rigid global regulation of the internet, undermining the dominance of American internet companies, stifling innovation and restricting the openness that underpins the net. Perhaps America also fears that its own cyberwar effort has the most to lose if its well-regarded cyberspies and cyber-warriors are reined in.'

Cyberspace is now the 'fifth domain' of warfare. The virtual battleground has joined the killing fields of land, sea, air and space. The machines and the networks are being conscripted to the service of the ancient and incurable human habit of warfare.

Technology has always, to a depressing extent, been driven by the demands of the military. We are already used to the idea of killing people with remotely controlled, pilotless aircraft –

drones – a technology that further isolates the killer from the killed, a moral development in that it removes any sense of the individuality of the victim and turns real battles into versions of computer games. Soon, we will have to grow used to the idea of cyberspace attacks on the systems on which we have come to rely.

At first glance, attacks on computers and networks may seem physically less harmful, more a case of inconvenience than bloodshed. But, in fact, cyberattacks can change the physical world.

The Stuxnet computer 'worm' that emerged in 2010 is designed to make machinery go dangerously wrong. The Iranian nuclear programme was infected with Stuxnet. Operational capability at the Natanz enrichment plant was reduced by 30 per cent and a serious accident happened there in 2009. It is not, in this context, hard to imagine computer-controlled weapons being turned against their users by even more sophisticated software – or 'malware' as it is clumsily but poetically called. Fortunately, the command chain of America's nuclear deterrent is not linked to the public internet. On the other hand, Iran now boasts of possessing the second largest cyberarmy in the world.

This physical power of a cyberattack became apparent as long ago as 1982 when a gas pipeline exploded in Siberia because of a computer control system that the Russians had stolen from Canada. They had not realised – they would have had no way of knowing – that the CIA had rigged the system to malfunction and destroy machinery.

Estonia and Georgia both succumbed to cyberattacks in 2007 and 2008 originating, it is generally assumed, from Russia. The attack on Estonia is now known, ominously, as Web War 1. China is thought routinely to use cyberattacks to steal Western technology. Technical details of the new F-35, the Lightning II joint strike fighter, are thought to have been cyber-stolen by the Chinese. The American military now has a Cyber Command. The primary fear is military and strategic – cyberespionage is said to be the 'biggest intelligence disaster' since the loss of the secrets of nuclear weapons to the Russians in the late 1940s. But there is also crime. President Obama has suggested that $1 trillion was lost to cybercrime in 2009, though the figure is disputed.

Dealing with such weapons diplomatically is almost impossible. NATO is debating whether a cyberattack on one of its members is sufficient to count as an attack on all and thus justify a multinational response. Satellites can watch missile bases or airfields, but are blind to computer programmers. States can easily deny they were responsible for attacks originating from their territory, even if the geographical source can be located. Russia has called for a treaty, a move welcomed by General Keith Alexander of Cyber Command, but, in the present state of the technology, it is hard to see how it could be enforced.

Future possibilities are almost nuclear in their scale. A concerted attack could bring down military command, oil refineries, air-traffic control systems, electricity grids and transportation, effectively destroying a country's ability to function. Or there could be an attack on the internet itself, probably by cutting the undersea fibre-optic cable network which carries 90 per cent of web traffic and which is dangerously concentrated in a few critical locations.

The threat of cyberwarfare is real but incalculable. It is evidence of the growing fragility that arises from a fully interconnected system. Just as the banks failed because they had all, in effect, become one giant bank, so nations could fail because, in cyberspace, they have all become one nation. The old, complex multiplicity at least had, until the advent of nuclear weapons, the virtue of limiting warfare roughly to the territory of the combatants.

The idea of controlling cyberspace in the name of peace is a challenge to the libertarian geek ideology which speaks with the loudest voice on the internet. Cyberspace must be free, they say, because that is how it will liberate the world, tearing down frontiers and uniting people. We will become ethical creatures through the medium of the free internet. Any opponent of any aspect of this view will be accused of being a Luddite, the contemporary equivalent of the nineteenth-century wreckers of textile machinery who feared the destruction of working-class jobs.

But the possibility of a highly destructive cyberwar makes it clear that the simple contest of geeks versus Luddites is futile. It is time for cyberethics, long considered a somewhat fusty discipline pursued by people who seemed out of touch with cyberreality, to enter the mainstream.

Technology extends human capabilities. What it extends and how is an expression of human choices which, in turn, are an expression of human nature. There is, however, a view that our current technologies, especially the phenomenon of global and permanent connectivity, may actually reform human nature, making us a more peaceable, empathetic species. This is the view both of hard-line cyberutopians – who tend to think all our new competences are, almost by definition, leading us to a better world – but it is also the view of some softer, liberal optimists. But is it, in the face of human history, realistic?

History, for the most enthusiastic cyberoptimists, is a timeline of technologies – printing, photography, telephony, rail and air travel – which leads inexorably to the smartphone and the iPad and they will, in turn, lead to a freer, richer, better world. The omissions from this timeline are obvious. After Gutenberg introduced moveable type into Europe in the mid-fifteenth century, the continent entered the bloody era of the wars of religion. The great transport and communication innovations of the nineteenth century paved the way for the twentieth, the bloodiest in human history. Technology may have made the world – some of it – richer, but there is no evidence from history that it has made the world – any of it – better.

The argument that the internet and mobile communications brings down barriers, crosses frontiers and subverts oppression feels persuasive. The 2011 uprisings across the Arab world (see Chapter Eight) did seem to be inspired and organised by the ability of the protesters to see what was happening elsewhere, to publicise the violence of the authorities and to arrange their demonstrations.

Yet there are also countless ways in which governments in countries like Russia, China and Iran have proved just as adept as using these technologies as their opponents. Does the instrument of liberation become another tool for the tyrant?

There are subtler tyrants than men in army uniforms and sunglasses. This book is also about the human ability to survive another, more insidious and seductive threat than warfare or explicit political oppression. This threat comes from our eagerness – most clearly expressed in the cultish idea of the Singularity – to unite with or change ourselves into machines, to simplify ourselves into becoming machine-readable and machine readers. It is another short-term solution to a long-term problem – that of being human. Cyberethics should find a way of exposing and discussing our immoderate love of the machine.

The awareness that this is a threat has been growing steadily, particularly among the technocrats. Pattie Maes is a Belgian computer scientist working at MIT. Her speciality is human–computer interaction and she has, in development, a startling contraption consisting of a combined smartphone, video camera and projector. The wearer of this can instantly call up information about anything he looks at. He can even draw a circle on his wrist and cause a watch face to be projected. Facial recognition technology will enable information about people he meets to be projected on to their bodies. Maes is, in other words, at the leading edge of gadgetry. But even she has doubts. 'As designers of tools and products,' she says, 'and technologies we should think more about these issues.'

Maes argues that gadget designers should take on the ethical issues of their products, especially when human skills are being successfully simulated by machines. Cyberethics requires, above all, a definition of the human which will prevent that realm being invaded by the machine.

Jaron Lanier, the Silicon Valley apostate whose views were discussed earlier, embraces a modified Luddism – 'an explicit social contract between the engineers and society to create not just jobs but better jobs'.

Sherry Turkle, also of MIT, calls for a step back from technological determinism in favour of 'realtechnik', an honest assessment of what we really want from the machines. In her work, especially with children, she has observed the way networks obstruct rather than encourage human contact. 'At the extreme,' she writes, 'we are so enmeshed in our connections that we neglect each other.'

These are dissidents from the cyberutopian vision. For them, our new machines should be judged ethically. We require tools of human flourishing, not passively accepted technologies to the logic of which we must consign our futures. The future is shrinking says Danny Hillis of the Long Now Foundation. This may be because we have handed it over to the geeks and their gadgets.

But cyberethics is a new and callow discipline with little rhetorical or persuasive power in the face of the market. At the time of writing, four of the top ten biggest companies in America by market capitalisation – Apple, Microsoft, IBM and Google – are in the business of selling the gadgets and networks of cyberutopian dreams. Apart from Google's motto 'Don't be evil' and Apple's rather fussy sexual censorship of its apps, there is little evidence of a communal corporate ethos other than the commitment to new technology as unarguably good – or 'insanely great' as Steve Jobs routinely and slightly disturbingly describes Apple products. But perhaps ethics is not their business; perhaps it is ours.

We, however, are plural and the new technologies are sold to many countries with many different value systems. Even within Britain and the US there is very little sense of shared values. Aaron Sorkin told me in the course of our interview that there were two distinct audience responses to the rise of the supreme geek Mark Zuckerberg. Under-thirty-fives tended to see it as a heroic tale of a young man's vision overcoming all obstacles. Over-thirty-fives saw it as a morality tale about Zuckerberg's betrayal of his friends in the pursuit of power. Even if the young do have doubts – Sherry Turkle gives many moving examples of children who feel bullied, threatened and exhausted by the hyper-connectivity of their lives – it is almost impossible for them to stand back and exercise judgement about the virtual societies they are now required to inhabit.

If children are trapped, adults are seduced. When the iPad was released, middle-aged men and women at once appeared swaggering about with them like teenagers with a new pair of trainers. I have been, in my time, a swaggerer with new gadgets; I retain the impulses of the early adopter who met Bill Gates in 1994. Lately the iPad adopters have ceased to swagger and started stroking and hugging them to their chests. The average age of computer gamers, meanwhile, has risen to the mid-thirties and Turkle's young interviewees are at their most alienated when describing their parents' fixations on their BlackBerries. She tells the story of sixteen-year-old Audrey. When she is picked up from school not a word is exchanged. Her mother looks up briefly when she opens the car door but then returns to concentrating on her phone. Sometimes they do not converse at all for four days. 'It [the phone] gets between us,' says Audrey, 'but it's hopeless. She's not going to give it up.'

Cyberethics for adults also faces a challenge that goes beyond consumer gadgetry. There are the spectaculars organised by the technology labs – spectaculars like Watson. In February 2011 Watson, an IBM computer named after the company's founder, Thomas J. Watson, played a two-game match of the TV quiz game *Jeopardy!*. It was playing against Brad Rutter, the show's biggest money winner, and Ken Jennings, the record holder for the longest winning streak.

At the most superficial level, this was a huge public relations coup for IBM. First, Watson won by an enormous margin and it won at something very familiar to millions of people – *Jeopardy!* has been on US television since 1964. Secondly, people enthusiastically embraced the idea that it was a thinking machine. This was in spite of the fact that Watson was represented by a black, curiously tombstone-like slab decorated only with a globe surrounded by dashes suggestive of light or, possibly, hair, and criss-crossed by animated orbiting streaks. It could have been the alien monolith from *2001: A Space Odyssey*.

Even Watson's mistakes generated a warm, comic mood. *Jeopardy!* is a kind of reverse quiz in which an answer is given and the contestants have to come up with the right question. One round involved coming up with the question in the category US Cities, to which the clue was: 'Its largest airport is named for a World War II hero; its second largest, for a World War II battle.' The two humans got the correct answer – 'What is Chicago?' Watson mystifyingly replied, 'What is Toronto?', though the five question marks on the screen indicated it was not very confident of this answer. The audience gasped but then laughed and applauded warmly when it turned out that Watson, though wrong, had been cautious enough to bet a very small amount of its money on Toronto.

But the PR coup was based on a genuine technological coup. In 1997 another IBM machine, Deep Blue, beat the greatest of all human chess players, Garry Kasparov. It was a fairly hollow victory. Though chess has long been considered the clever people's game, it is, in fact, very simple compared to *Jeopardy!*. All Deep Blue did was assemble and then search the accumulated wisdom of the chess players involved in its programming. Also it could scan future board states after many more moves than any human player. Its match-winning advantage was that it could do all this very quickly and accurately. Deep Blue also did not respond to Kasparov's notoriously aggressive presence in the way that human players did – with fear.

Watson, in contrast, was required not just to scan its memory and make forecasts within the strict parameters of an 8 x 8 chessboard; it was required to understand 'natural', i.e. human, language and to provide itself with exact probability assessments of the rightness of its answers. This was impressive. Machine understanding of spoken human language has proved night-marishly difficult to achieve – as the ponderous computerised call tree inquisitors so often demonstrate. Indeed, the problem has acquired a special geek name – it is known as the 'Paris Hilton problem'.

This particular blonde celebrity was born into the family that founded the Hilton Hotel chain. Her parents, mysteriously and perhaps insensitively, gave her the name of the capital of France and thereby made her sound like one of their properties. If they had not taken that step, Paris Hilton would still only denote buildings on the rue de Courcelles or in the Place de la Défense. But she became so famous that, for human beings at least, the term only evokes a blonde with a prison record and a fashion addiction.

For computers, however, the confusion remains. Getting a machine to understand in which sense the phrase is being used is extremely difficult. Humans know at once, but it is hard to explain to a machine how they know. They know through the infinite complexities of context: the expression on the speaker's face, the tone of the voice, the shape of the sentence, the sort of person who is speaking and where they are speaking. A machine that understands such contextual complexities does not yet exist.

But Watson did show that IBM had solved a problem that had dogged AI research for decades. It had demonstrated the possibility of a genuine 'expert' system that could, for example, perform medical diagnoses or analyse and personalise investment strategies. Or Watson might help out Facebook by mining the data of its six hundred million users and novel connections, all of which could be used in marketing. Watson also has the ability to learn from its mistakes through a system of feedback loops. This meant that it could, in time, become smarter than its programmer as opposed to Deep Blue, which was just quicker.

There was also a sense in which Watson was introspective. As it gave its answers, a bar appeared at the bottom of the screen indicating, in percentages, its confidence in its answers. This was a statistical calculation based on the number and quality of its sources. But there was a ghostly sense that this might foreshadow a machine that can do something that, so far, only we can do – think about our thinking.

Nevertheless, Watson is still a long way from human equivalence. It may be able to improve itself as a question-answering machine, but it cannot decide to do something else; it is utterly devoid of creativity. All it has done, in fact, is solve some computing problems that have been around ever since AI was first conceived in 1956. After the contest, sceptics pointed this out, some very grumpily. 'I'm not impressed by a bigger steamroller,' said Noam Chomsky, the great linguist-turned-philosopher and cognitive scientist. 'Watson understands nothing,' he added, 'it's a bigger steamroller. Actually, I work in AI, and a lot of what is done impresses me, but not these devices to sell computers.'

Sceptical neuroscientist, journalist and author Jonah Lehrer, meanwhile, drew attention to one glaring but otherwise unnoticed deficiency in Watson's capabilities: it is very inefficient. Lehrer pointed out that the human brain does what it does using power equivalent to a 12-watt light bulb. He does not know the power consumption for the much less impressive Watson but estimated it would be tens of thousands of times more. In other words, it still requires almost industrial age engineering for the computer to understand a little English.

Computing and management thinker John Seely Brown – commonly known as JSB – takes the sceptical position a little further by asking not what Watson is but what humans are. 'The essence of being human,' he says, 'involves asking questions, not answering them.'

JSB has a point with this distinction. Asking questions is a creative act; answering them need not be because the answer is usually predetermined – or even created – by the question. This may be why in my brain scan I found it hard to lie in response to direct questions but easy when I was making up a story. The false stories were my creation, I was asking the questions and giving the answers, but the questions were an external, moral force which was, in the heat of the moment, defined by the fMRI machine, hard to resist with a lie.

Humans are questioning so they make machines that can give them straight answers – or questions in the perverse rules of *Jeopardy!* – like Watson. Humans, being complex and context-sensitive, are less good at straight answers. This can inspire sentimentality about the simple-minded machine. In the movie *2010*, the sequel to *2001*, it is revealed that what turned the computer HAL into a murderer was the pressure of keeping a secret. 'HAL was told to lie,' says Dr Chandra, the machine's designer, 'by people who find it easy to lie. HAL doesn't know how.'

Chandra is being a sentimental, deluded geek. HAL was, in truth, mendacious in the extreme – notably when he told the crew in the first movie that there was a malfunction when there wasn't, the plot device that was the mainspring of both movies – but that is because HAL was creative like a human being. He asked questions of himself and came up with rational but homicidal answers. He was, in other words, introspective, a quality just about detectable in Watson in the form of its calculations of the likelihood of its answers being right.

But will some combination of what we learn from the neuroscientist's scanners and IBM's supercomputers eventually crack all these problems and produce a thinking, introspective, creative machine, a machine that can come up with complex solutions for a complex world? Nobody – neither geeks nor Luddites – can currently give an honest answer to that question. We can, however, ask what it would mean.

Transhumanists and fans of the Singularity would say it would mean we were at last able to transcend the limitations of our biological condition – the limitations of death and disease as well as war and all the other follies handed down to us by evolution. But, of course, any such machine would have humanity and all our follies and virtues built into it – we would, after all, be its makers. On the other hand, it could choose to go beyond our limitations to something better. But what would that be? Either the machine is human or inhuman – if the latter then it will not be us who have transcended our condition, but the machine. A new species might thus emerge, but then it would have no meaning for us.

That may be science fiction, but, as I have shown, something like it is the accepted orthodoxy, the ideology, of our age. It is certainly involved in the marketing of gadgets that extend our capabilities even as they seek out and sell our identities. And the marketing works; we plainly want to give ourselves to these gadgets. But what would we lose?

I have more than fifty original David Hockney pictures on my iPhone. One arrives every few days, usually early in the morning. I look at each in wonder and come back to them repeatedly. They are, for the most part, restful, ordinary scenes. There are views from Hockney's bedroom window, bowls of oranges and apples, a pair of shoes, vases of flowers and several lovingly drawn glass ashtrays – Hockney is an enthusiastic smoker and an angry campaigner against all smoking bans. It is strange, miraculous and consoling to receive these lovely things with such immediacy from one of the great artists of our time. He emails his pictures to a small and privileged mailing list the moment he has made them, often before he has got out of bed.

In 1994, when I met him, even Bill Gates could not have dreamed of the technology that makes this possible. Hockney 'paints' on an iPad using an application called Brushes. The application provides an infinity of colours, variable transparency and brush thickness and an amazingly sensitive and paint-like response to the touch of a finger. In fact, the effect feels so close to paint that Hockney found himself wiping his finger on his trousers or his jacket every time he wanted to change colour.

Years ago he started having Saville Row suits made with an especially large inside pocket to accommodate his sketch books. Now the pockets only carry his iPad; they are, happily, just the right size. The machine has replaced paper, pencils and brushes. It has, Hockney points out, some clear superiorities. 'It's a luminous medium,' he told me, 'and luminous subjects suddenly become very interesting – light, light hitting glass, things that are shiny.

'You can have as many layers as you like. With watercolours you can't have more than three layers or it turns muddy. It's not a surface so you don't damage it. And it's like an endless piece of paper, you can begin at one size and you can extend or reduce the drawing. It's like sticking extra pieces of paper on the edge. It takes you a while to grasp you can start drawing on scales you couldn't think of before.'

The iPad has one extraordinary feature that seems, in some way, to change our entire perception of the image. It can retrace the entire drawing process as an animation so that every touch of the artist can be seen. 'Normally you don't watch yourself draw when you are drawing because you are always a few moves ahead at any given moment, you will always be thinking of the next mark, the next mark ... But with this you can watch it all as it happened. The only thing like it is that great Picasso film where he's drawing and he realises it wasn't the finished drawing he was doing that was the subject, it was what he was doing.'

There was one way in which, for Hockney, the meticulous draughtsman, the iPad was not as good as a sketch book. The finger is a very thick drawing implement. This meant, for example, that he could not be sure he had drawn thin lines so that they met exactly where he wanted them to. Now he uses a special rubber-tipped stylus.

In the 1960s and 1970s, Hockney's paintings such as *A Bigger Splash* and *Mr and Mrs Clark and Percy* were masterpieces that captured the sunlit strangeness of a moment in time. They were 'pop' art certainly but also timeless and monumental. Ever since, he has walked the familiar tightrope of the artist, suspended between the demands of topical and of the timeless.

He left England for Los Angeles in 1978. He was born in Bradford in West Yorkshire, but, while he was in LA, the family had begun moving eastwards. Margaret, his sister, arrived in Bridlington on the East Yorkshire coast, and was joined by his mother in 1989. Hockney bought them a grand, double-fronted and oddly styled house just off the seafront. He says it was built in 1924 by a trawlerman 'for his ugly daughter.' In the 1990s he travelled back and forth to Yorkshire to visit a dying friend and, back in California, started painting the county's landscape from memory. His mother died in 1999 and, six years later, Hockney moved into the house, while maintaining a place and staff in LA. Bridlington is a seaside resort stuck in the Fifties and few people travel to East Yorkshire for the quality of life or the tourist attractions. It is, in short, a very quiet place. This means, to Hockney's immense satisfaction, that he is left alone with his assistants to get on with his work.

He works on the top floor of the house on giant eighteen-screen video works, primarily of a nearby tree-lined lane he loves. He believes these works with their multiple points of view and varied timings reflect more accurately than 3D movies what the human eye sees and the brain experiences.

'Yes, you can make TV pictures all in 3D. It's good for pornography, for big tits and asses, but not much else. It's still one point of view. It's not how the human eye sees ... Actually seeing eighteen separate films is more like human vision than one camera could ever be. All that photography has ever done for us is make us claustrophobic.'

He also works in a 10,000-square-foot warehouse around in which he spins on any one of four wheelchairs. He does not need them; he just likes them as modes of transport. Most people would be squeamish about using wheelchairs like this, but he is a man who was always defiantly himself, never more so than now, at seventy-four, back in Yorkshire.

Paintings and drawings were soon accompanied by photo-collages – he called them 'joiners' – made out of dozens of Polaroid photos. He had become obsessed with the way the camera had changed our way of seeing and the joiners were attempts to use the machine to overcome its own primary shortcoming – its immobile, one-eyed perspective.

With the physicist Charles Falco, he developed the theory that cameras had, in fact, been in use long before the official date of the invention of photography in the 1820s. The camera in the form of the *camera obscura* – a darkened room into which images of the exterior are projected – had been around for two thousand years. With the invention of lenses, this became ever more effective; the theory was that the device was used by the Old Masters from the Renaissance onwards. The Hockney–Falco theory has run into criticism, especially from art historians who think it implies the Old Masters cheated, but he is convinced. 'I'm positive it's true. There is no question. You can compare the way Caravaggio used optics about 1605 to the way they were used by Vermeer sixty years later. That was a fascinating period because lenses were developing enormously because of the microscope. By the time you get to Vermeer the lenses were probably quite close to what we might have in a camera today.'

Hockney's message is clear. Technology and the image – the machine and the hand – are natural companions. In taking up the iPad, just as when he took up the Polaroid camera, he is following an ancient tradition of innovation in the visual arts.

Before I saw him I had been to an exhibition of his work on iPad and iPhone in Paris. It may have been the first exhibition that could have been emailed to the gallery. On one wall were twenty iPhones arranged in rows, on another a horizontal line of twenty iPads. The pictures on each changed continuously. In another room there were projected images of the pictures, also constantly changing.

Occasionally a drawn message will flash up on the screens.

'Made for the screen,' says one, 'Totally on the screen. It's not an illusion.'

Or, most pointedly:

It is thought new technology is taking away the hand. I'm not so sure. If you look around a lot is opening up. Love, David H.

By some electronic conjury, one of his assistants can get files out of his iPad large enough to print large versions of the pictures. He could actually sell these; the emailed versions were being given away free. But at the exhibition, where the darkened rooms made his electronic pictures look like the most glorious stained glass, I realised that the free iPad versions, generously emailed to his friends, were the true originals. The buyer of the pricey print would just be getting a copy.

Paul Cézanne said of Claude Monet that he was 'only an eye, but my God, what an eye!'. You could say the same of Hockney. But there is something more. There is his fascination with the technology, with the pros and cons of these new machines from the *camera obscura* to the iPad. He engages with them, neither entranced, like the geek, nor mistrusting, like the Luddite.

Art, the highest expression of the human mind, is the complex solution to the complex problem of our existence, and Hockney, stroking the screen of his iPad, is the best possible vision of the human future – the artist at peace with the machine, his servant.

EPILOGUE

Six months into researching this book I came across a reference to a competition to design a computer game based on the character of Emily Dickinson. Taking part were three heroes of the gaming world, Will Wright, the creator of *Sims* and *Spore*, Peter Molyneux, designer of *Black & White*, and Clint Hocking, designer of *Splinter Cell*.

It was a lighthearted event and reported as such in the gadget press. Wright won with the idea of a highly interactive game stored on a USB flash drive. The character with which the player interacts was Dickinson.

'As you interact with her, you start with a cordial relationship,' Wright explained; 'she becomes romantically obsessed with you, or goes into a suicidal depression, and at the end, she can delete herself from the memory stick.'

Happily, the game was never actually made.

ACKNOWLEDGEMENTS

I would like to thank the following for discussions specifically related to this book: Nigel Biggar, Michael Burleigh, Adam Curtis, Paul Davies, Jerry Fodor, John Gray, John Horgan, Jaron Lanier, Steve Lansing, James Lovelock, Iain McGilchrist, Paul Ormerod, Larry Parsons, Scott Patterson, Roger Scruton, Dan Simons, Robert Rowland Smith, Dave Snowden, Nassim Nicholas Taleb, Mark Vernon and Paul Wilmott.

Through journalism I have, over many years, met or spoken to others whose thoughts and work have influenced the arguments and impressions in this book. There are too many to list, but I would like to mention the following: Peter Ackroyd, Nancy Andreasen, Marc Andreesen, Jacopo Annese, John Ashbery, Jeff Bezos, Paul Bloom, Stewart Brand, James Cameron, Richard Dawkins, Jared Diamond, Bill Gates, James Gleick, Steve Grand, Michael Haneke, Stephen Hawking, Perez Hilton, David Hockney, Jonathan Huebner, Michael Landy, Brenda Milner, Shigeru Miyamoto, Christopher Ricks, Marilynne Robinson, Burt Rutan, Aaron Sorkin, Steven Spielberg, Tom Stoppard, Arthur de Vany, Craig Venter and Will Wright.

In addition, I would like to thank the publications which gave me the chance to meet these people or just to think about these matters. Primarily, there is the *Sunday Times*, but also the *Times*, *Independent* and *New Statesman*. I am grateful to *Vanity Fair* for giving me the chance to look more closely at the story of Henry Molaison.

Acknowledgements

Thanks also to my agent, David Miller, and, finally, to Christena, my wife, who helped in more ways than I can possibly list here.

FURTHER READING

This is a highly selective list of books that I refer to or have influenced *The Brain is Wider than the Sky*.

INTRODUCTION

John Ashbery, *Wakefulness* (Carcanet Press, Manchester, 1998)

Lyndall Gordon, *Lives Like Loaded Guns: Emily Dickinson and Her Family's Feud* (Virago, London, 2010)

Stuart Kauffman, *Reinventing the Sacred: A New View of Science, Reason and Religion* (Basic Books, New York, 2008)

This is the full text of the Emily Dickinson poem:

> The Brain – is wider than the Sky –
> For – put them side by side –
> The one the other will contain
> With ease – and You – beside –
>
> The Brain is deeper than the sea –
> For – hold them – Blue to Blue –
> The one the other will absorb –
> As Sponges – Buckets – do –
>
> The Brain is just the weight of God –
> For – Heft them – Pound for Pound –
> And they will differ – if they do –
> As Syllable from Sound –

Further Reading

CHAPTER 1: A DIVIDED MAN

Bill Gates, *The Road Ahead* (Viking, New York, 1995)

CHAPTER 3: HENRY

Antonio Damasio, *Descartes' Error: Emotion, Reason and the Human Brain* (Vintage, London, 2006)

John Gray, *The Immortalization Commission: The Strange Quest to Cheat Death* (Allen Lane, London, 2011)

Philip J. Hilts, *Memory's Ghost: The Nature of Memory and the Strange Tale of Mr M.* (Touchstone, New York, 1996)

Marvin Minsky, *The Emotion Machine: Commonsense Thinking, Artificial Intelligence and the Future of the Human Mind* (Simon & Schuster, New York, 2007)

CHAPTER 4: A SIGNATURE SCIENCE

John Horgan, *The Undiscovered Mind: How the Brain Defies Explanation* (Weidenfeld & Nicolson, London, 1999)

Julian Jaynes, *The Origins of Consciousness in the Breakdown of the Bicameral Mind* (Houghton Mifflin, New York, 1976)

Iain McGilchrist, *The Master and his Emissary: The Divided Brain and the Making of the Western World* (Yale University Press, New Haven, 2009)

CHAPTER 5: COUNTDOWN TO THE SINGULARITY

Jerry Fodor, *The Mind Doesn't Work That Way: The Scope and Limits of Computational Psychology* (MIT, Cambridge, 2001)

Ray Kurzweil, *The Singularity is Near: When Humans Transcend Biology* (Viking, New York, 2005)

CHAPTER 6: HITTING ZERO

Rodney A. Brooks, *Flesh and Machine: How Robots Will Change Us* (Vintage, New York, 2002)

Alan Deutschman, *The Second Coming of Steve Jobs* (Broadway, New York, 2001)

Bill McKibben, *The End of Nature* (Random House, New York, 2006)

Kevin Warwick, *I, Cyborg* (Century, London, 2002)

CHAPTER 7: MEN WITHOUT CHESTS

James Gleick, *The Information: A History, a Theory, a Flood* (Pantheon, New York, 2011)

N. Katherine Hayles, *How We Became Posthuman: Virtual Bodies in Cybernetics, Literature and Informatics* (University of Chicago Press, Chicago, 1999)

C. S. Lewis, *The Abolition of Man* (Zondervan, Grand Rapids, 2001)

Hans Moravec, *Robot: Mere Machine to Transcendent Mind* (Oxford University Press, New York, 2000)

Norbert Wiener, *Cybernetics or Control and Communication in the Animal and Machine* (MIT, Cambridge, 1962)

CHAPTER 8: PIMP MY ULTRASOUND

Jaron Lanier, *You Are Not a Gadget: A Manifesto* (Allen Lane, London, 2010)

Eli Pariser, *The Filter Bubble: What the Internet is Hiding from You* (Viking, New York, 2011)

Clay Shirky, *Here Comes Everybody: The Power of Organizing without Organizations* (Penguin, New York, 2009)

James Surowiecki, *The Wisdom of Crowds* (Anchor, New York, 2005)

CHAPTER 9: IT'S ALL IN THE GAME

Tom Chatfield, *Fun Inc.: Why Games Are the 21st Century's Most Serious Business* (Virgin Books, London, 2010)

Nicholas Christakis and James Fowler, *Connected: The Amazing*

Power of Social Networks and How They Shape Our Lives (HarperPress, London, 2010)

Paul Davies, *The Eerie Silence: Are We Alone in the Universe?* (Allen Lane, London, 2010)

Steve Grand, *Growing Up With Lucy: How to Build an Android in Twenty Easy Steps* (Weidenfeld & Nicolson, London, 2004)

Sherry Turkle, *Alone Together: Why We Expect More from Technology and Less From Each Other* (Basic Books, New York, 2011)

Gabe Zichermann and Joselin Linder, *Game-Based Marketing: Inspire Customer Loyalty Through Rewards, Challenges and Contests* (Wiley, New York, 2010)

CHAPTER 10: MONEY MEN

Adam LeBor, *The Believers: How America Fell for Bernard Madoff's $65 Billion Investment Scam* (Weidenfeld & Nicolson, London, 2009)

Paul Ormerod, *Why Most Things Fail: Evolution, Extinction and Economics* (Faber & Faber, London, 2005)

Scott Patterson, *The Quants: How a New Breed of Math Whizzes Conquered Wall Street and Nearly Destroyed It* (Random House, New York, 2010)

Nassim Nicholas Taleb, *The Black Swan: The Impact of the Highly Improbable* (Random House, New York, 2007)

Arthur de Vany, *The New Evolution Diet: What Our Paleolithic Ancestors Can Teach Us about Weight Loss, Fitness and Aging* (Rodale, New York, 2010)

CHAPTER 11: NOW NEUROAESTHETICS

Paul Bloom, *How Pleasure Works: The New Science of Why We Like What We Like* (W. W. Norton, New York, 2010)

Daniel Levitin, *This is Your Brain on Music: The Science of a Human Obsession* (Penguin, New York, 2007)

V. S. Ramachandran, *The Tell-Tale Brain: A Neuroscientist's Quest for What Makes Us Human,* (W. W. Norton, New York, 2011)

Oliver Sacks, *Musicophilia: Tales of Music and the Brain* (Picador, London, 2008)

Semir Zeki, *Splendors and Miseries of the Brain: Love, Creativity and the Quest for Human Happiness* (Wiley-Blackwell, London, 2008)

CHAPTER 12: THE NEW FOUND LAND

Michael Fitzgerald, *The Genesis of Artistic Creativity: Asperger's Syndrome and the Arts* (Jessica Kingsley, London, 2005)

Jacques Hadamard, *The Psychology of Invention in the Mathematical Field* (Dover, New York, 1954)

Michael Landy, *Everything Must Go* (Ridinghouse, London, 2008)

Marilynne Robinson, *Absence of Mind: The Dispelling of Inwardness from the Modern Myth of the Self* (Yale University Press, New Haven, 2011)

Marilynne Robinson, *The Death of Adam: Essays on Modern Thought* (Picador, London, 2005)

CHAPTER 13: THE AGE OF COMPLEXITY

Stewart Brand, *Whole Earth Discipline: An Ecopragmatist Manifesto* (Viking, New York, 2009)

James Lovelock, *The Revenge of Gaia: Why the Earth is Fighting Back and How We Can Still Save Humanity* (Allen Lane, London, 2006)

CHAPTER 14: THE PARIS HILTON PROBLEM

David Hockney, *Secret Knowledge: Rediscovering the Lost Techniques of the Old Masters* (Thames & Hudson, London, 2006)

Jonah Lehrer, *The Decisive Moment: How the Brain Makes up its Mind* (Canongate, London, 2009)

INDEX

Index

Index